Critter Tales

Also by Leigh Tate

IN PRINT

5 Acres & A Dream The Book:
   The Challenges of Establishing a Self-Sufficient Homestead

E-BOOKS

The Little Series of Homestead How-Tos
   How To Garden For Goats: gardening, foraging, small-scale
      grain and hay, & more
   How To Preserve Eggs: freezing, pickling, dehydrating, larding,
      water glassing, & more
   How To Make a Buck Rag: & other good things to know about
      breeding your goats
   How To Make an Herbal Salve: an introduction to salves,
      creams, ointments, & more
   How To Mix Feed Rations With The Pearson Square: grains,
      protein, calcium, phosphorous, balance, & more
   How-To Home Soil Tests: 10 DIY tests for texture, pH,
      drainage, earthworms & more
   How To Make Mozzarella from Goats Milk:
      plus what to do with all that whey including make ricotta

Critter Tales

What my homestead critters have taught me about themselves, their world, and how to be a part of it

Leigh Tate

Kikobian Books
www.kikobian.com

A necessary disclaimer: The purpose of this book is to entertain and encourage others with interests similar to me, the author. Within its chapters you will find accounts of my diagnosing and treating my animals for a variety of problems. Please note that I am not a veterinarian nor an expert in this area. You will not find enough information to diagnose and treat your own animals, you will merely find a narrative of my experience, including the use of herbs in treatments. Using herbs for medicinal purposes is highly controversial and requires a strict admonition to seek professional medical advice first. If your animals exhibit any of the same symptoms, it is *your* responsibility to find the correct diagnosis and treatment. This should be done through a veterinarian. These tales are not an endorsement of my methods, but simply a collection of things that happened and what I did. There is no guarantee you would have the same results, so please use appropriate cautionary protocol in this area.

This book was built from the ground up with open source software on an open source operating system: Ubuntu Linux 12.04 (Precise Pangolin), Xfce 4.10 Desktop Environment, LibreOffice Writer 3.5.7.2 word processor, Zim 0.54 desktop wiki, Notes 1.7.7 quick notes plugin, gedit 3.4.1 text editor, Gimp 2.8 photo editor, and Scribus desktop publisher versions 1.4 and 1.5. Also used were open source fonts Kaushan Script, EB Garamond, EB Garamond SC (all SIL Open Font License Version 1.1) and Liberation Sans (GNU General Public License v.2).

ISBN 978-0-9897111-2-8

Kikobian Books
www.kikobian.com

*To Dan, my partner in critter-keeping,
homesteading, and all of life.*

CONTENTS

ACKNOWLEDGEMENTS

The creation of a book is a team project. As the author I get to have my name on the cover, but getting it from an idea in my head to the book that you now hold in your hands required help.

The heart of any book is the words which express the author's ideas and communicate them to you, the reader. My beta readers read through the earliest drafts of *Critter Tales* and gave me valuable feedback on content, continuity, and clarity. Their questions and suggestions did much to shape the telling of the tales. To Garrett Alley, Sharon Campbell, Anna Hess, Nathan Huntley, Perry Overton, Elizabeth Sweet, and Barbara Wills, a heartfelt thank-you.

Once the ideas are committed to words, the book itself is designed. It is the proofreaders who take the freshly formatted pages and search them for the little mistakes that can distract the reader from an enjoyable reading experience. My sincerest thanks to Heather East, Fern Feral, Carolyn Miller, and Ellen Leigh Sadler for doing just that. If there are any errors left, they are entirely my own.

A very special thank-you goes to Leslie Koster, my editor. Her enthusiasm, encouragement, and expertise have been a blessing to me every step of the way. It is my sincerest hope that this humble, self-published book is a credit to her.

The one to whom I am most grateful is my husband Dan. These are not just my tales but ours. He was always willing to discuss my ideas, gave me helpful feedback, and never complained when I was lost to the computer and late for chores.

CONCERNING CRITTERS

Critters are the center of our homestead world. From the time we get up, to the time we go to bed, even during the night, our life revolves around them and their needs. It is a partnership, really, which provides a framework for the functioning of our homestead and the stability to our lives. We are sometimes pitied for being "tied down" by our animals, and it's true that there can be no spontaneous trips or long absences for us. Not for both of us, anyway. But this is the life we choose, a committed relationship with the land and the lives it sustains.

My husband Dan and I come from typical middle class suburban backgrounds. Our experience with livestock was grounded in what we learned in kindergarten and from visiting petting zoos. We knew farm animals provided many useful things, such as food in the form of eggs, milk, and meat. As gardeners, we knew that manure is an important source of nitrogen for good compost, which is essential for growing healthy fruits and vegetables. As a handspinner, I knew all about fleece and fiber. When it came to choosing and caring for livestock, however, we knew we had a lot to learn.

We started our homesteading journey with a purchase of five acres. When we bought it, part of it was cleared and part of it was wooded. It had no fencing and no barn, but there were two old outbuildings that could be a beginning for animal housing. There was much uncertainty. What kind of fencing would we need? How many animals per acre could we keep? Would our outbuildings be adequate for shelter? Were they big enough? What about feed? We knew we would need to buy most of our feed in the beginning, but we wanted to work toward feeding all of our critters from what we could grow ourselves. What would we need to grow? How much? What would it take to process it and how would we store it?

We pondered what kinds of animals to get: chickens, turkeys, goats, a cow, a guard dog? What breeds? How many? Could they live together? Would they need their own areas? How big would those areas need to be? Our biggest concern was that we would not be able to care for our animals properly. Dan did not want our excitement and enthusiasm to overrule common sense. There were two considerations here - the critters themselves and the land. The last thing we wanted was to end up with more critters than we and the land could support.

The framework for all these questions was our homesteading goal of self-sufficiency. Many people associate this term with isolationism. For us, however, it had nothing to do with that. What we sought was a fuller experience of providing for ourselves. We wanted a stronger connection to the sources of our food and to be a part of life's seasonal rhythms. We both love being outdoors and working with our hands. We are not interested in the latest consumer trends or newest technological marvels. We sought to rely more on doing and less on buying. We would read *The Little House on the Prairie* books or Ralph Moody's *Little Britches* series and admire the characters' contentment with what others deem a hard way of life. The simplicity of their lifestyles appealed to us.

When we bought our little place in early 2009, we had big plans. Even so, we spent close to a year simply observing. We noted the direction from which the rain came and which ways the wind blew. We observed which areas received the most sun and which were slow to drain after a storm. We walked the land and discussed what we hoped to accomplish. Where would be the best spot for a garden? For an orchard? Where would we build a barn? How about beehives? We took notes, made sketches, and created a Master Plan to map out what we hoped our homestead would look like someday.

It was during this first year that I spent a lot of time researching various critters so that I could understand the various aspects of their care. Thanks to the internet, there is a lot of information available, but

I had to learn how to weigh each bit of advice against our situation and our goals. For example, I learned that alfalfa is a popular feed for goats, but I also learned that it does not grow well in our part of the country. If I wanted to grow our own hay I would need to look for an alternative crop. Another example is breeds of chickens. Living in the southeastern United States meant that I was less concerned with how winter hardy a breed is, and more concerned with how well they could tolerate our summer heat. I quickly figured out that there is no one-size-fits-all plan for either critter-keeping or homesteading.

When it came to getting animals, our plan was to start small and build to numbers which were manageable for us without overgrazing or overusing the land. One thing this meant was that we couldn't keep all the critters that were born or hatched on the homestead. We reckoned we would sell, trade, or eat the surplus. This purely rational decision points to a different kind of learning curve, one that involves the heart as well as the head. We not only had practical skills to learn, but emotional ones as well. Would we really be able to kill and eat animals that we had raised and come to know so well?

Critter Tales is the story of how we are learning the answers to all of these questions. It is divided into sections, one for each type of critter, with the sections divided into individual tales. With most sections I start with what I learned from my initial research. The tales themselves range from anecdotal to technical, and I'd have to say that we've learned just as much from our experiences as from my research. Without a livestock veterinarian in the area, most of our problem-solving relies heavily on that research. Although I am not trained in animal science or health, I will nonetheless share what I learned, what we did, and the outcomes, both good and bad.

Some of the sections are a chronological telling of our experiences, others are less so. In trying to share what we've learned, it made more sense to arrange much of the information topically—such as kidding problems or how to introduce new chicks to an existing flock—rather than chronologically. These aren't things we figured out the first go-round, but over the years. I felt my readers would benefit more from having all related information together in one tale, rather than spreading it out in bits and pieces throughout the various tales. I've tried to conclude each section by discussing the long-term impact each critter has had on our homesteading plans and goals, with a view to the future.

The hardest part of writing such a book is that the tales are never-ending. There is always another lesson learned, always something new to tell. At some point, however, a stopping place must be chosen so that the work of preparing the book for publication can commence.

Because some of the sections are prepared months in advance, I have indulged in a postscript to bring my readers up to date before the book must be finalized for the printing press. My most current critter tales can be read at my blog, www.5acresandadream.com, where the telling is ongoing.

Leigh Tate
November 2015

Chicken Tales

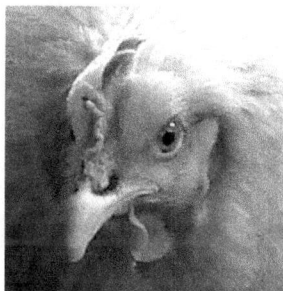

DOING MY HOMEWORK

Chickens were the first critters we decided to get for our homestead. It had been a long time since I'd had chickens, but even then my experience was primarily the gathering and cooking of eggs, although I'd helped with plucking feathers too. However, I'd never before been responsible for chickens. I'd never set up a coop, nor hatched them, nor raised them from chicks, nor maintained them for any length of time. During our first homestead winter I did some serious research on everything we would need to know to start our own little flock.

My first research notes looked something like this:

On average, 3 hens will give 2 eggs per day
Nest boxes: need 1 per 2 - 4 hens
Hens start laying at approximately 6 months of age
Peak egg production is the first year, decreasing afterward
Laying stops during molting
Laying stops when daylight falls below 14 hrs / day
Egg production is best at temps of 45° to 80°F (7° to 27°C)
If a rooster is desired, need 1 per 8 - 12 hens
Housing: need minimum of 4 sq. feet per bird for large breeds that are allowed outdoors

I also began to make a list of potential breeds, because they aren't all alike. I knew I wanted heritage rather than commercial breed chickens, but I made a checklist to help narrow down possible breed contenders. I wanted:

Dual purpose, for both eggs and meat
Relativity docile (read: not too noisy)
Good egg producers
Having brooding and mothering instincts

Rather than just choose one breed, however, I figured I'd get a batch of mixed breeds so we could test them out. In the end I chose Barred Hollands, Delawares, Ameraucanas, and Welsummers.

Being the analytical sort, the next thing I did was to set about trying to figure out how many chickens we would need. I considered two questions:

How many eggs do we want?
How many chickens can our land sustain?

Based on my research, I calculated that if we kept a dozen laying hens, we could probably count on about 8 eggs per day during peak production their first year. Put another way (according to the books), from 7 to 18 months of age I could expect an average hen to lay approximately 20 dozen eggs her first year and 16 to 18 dozen after her first molt. So, a dozen hens x 20 dozen eggs = 240 dozen eggs. I usually bought one or two dozen per week, so with a dozen hens we'd be using a lot more eggs that first year! Even at 16 dozen per year each after her molt, we were looking at about 192 eggs per chicken per year. That's still a lot of eggs. So said my calculations.

But did we have room for a dozen chickens? Our future coop was about 90 sq. ft. According to the books, large breed chickens allowed outside can be housed with a minimum of 4 sq. ft. per bird, which would mean we could keep 22 large birds. I thought my future dozen free-range chickens should fit comfortably in their new quarters.

Lastly I looked at a cost/benefit ratio for having chickens. Most folks do this at some point during their early days of chicken-keeping. While cost versus production wasn't my bottom line, it's hard not to analyze it this way, especially at first. I didn't just look at eggs, however. I also considered meat, manure, and more chickens. My calculations went something like this:

EGGS: Comparable to how I planned to raise my chickens, i.e. free-ranged, natural diet, etc., I could expect to pay about $3.50 per dozen locally. Since that price was beyond my egg budget, I definitely thought growing my own was worth it. I calculated that if I managed to get that 240 dozen eggs my first year, that's comparable to $840 worth of high-quality eggs. That's a lot for two people, but eggs can be

Eggs can be preserved in a variety of ways: freezing, dehydrating, and water glassing. (See "Resources" for where to find more information.) To freeze eggs, beat whole eggs as for scrambled, salt if desired, and pour into ice cube trays or muffin tins. Remove when frozen and store in a freezer bag or container.

beaten as for scrambled and stored in the deep freezer for about a year, so theoretically, homestead eggs can be available for use year-round.

Meat: At that time I priced comparable chicken for about $3.49 per pound. That would be for a whole, free-ranged chicken from local farmers who practice sustainable agriculture and who feed locally-grown grains. If a dressed chicken weighs approximately three pounds, then that would be $10.47 plus tax for a chicken. We can get two good meals from that, more if I make a chicken soup, stew, or casserole. In addition, there would be the bones to simmer with a little vinegar, lemon juice, or wine for a mineral-rich chicken broth.

Manure: Initially I had no idea of what chicken manure for the garden or compost pile would cost. I couldn't find it locally, but online it ranged from $3.50 to $9.70 per cubic foot for organic chicken manure, not including shipping. At those prices, I couldn't buy it for the size and number of gardens we planned to have. Considering that good compost is hard to make without manure, our own chicken manure would be worth its weight in gold!

MORE CHICKENS: Eventually I would need replacement chickens. If chicks cost $2 to $3 each plus shipping, then replacing them ourselves would be another cost savings.

COST: The initial cost would be the chicks, plus equipment, chick feed, and coop repair. After that I figured it would be mostly feed, whether I bought pre-mixed commercial, or made my own. Allowing chickens to forage and supplementing their diet with garden and kitchen scraps would help keep feed costs down. In fact, I might not even need the nutritionally-complete commercially-formulated ration for confined birds, especially because growing our own grains was (and is) the ultimate ideal and a goal we'd be working toward.

At that point, all of this was "book learnin'" or head knowledge. I had no practical experience. Going on five years later, I can tell you with great certainty that my chickens did not read the same books and websites as I did.

My first surprise was wintertime. Some folks leave a light bulb burning in their chicken coops during winter to increase egg production. We decided not to use artificial light in the coop. It just didn't seem natural. If chickens have natural cycles, we felt we should respect that. They stopped laying during their molt, but since I had four different breeds, they didn't all molt at the same time and I still

Our first chicks were mail order, but we hoped for home-brooded after that.

got eggs throughout the winter. My Delaware hen finished her molt and resumed laying before the shortest day of the year, along with my lone Barred Holland hen and my two Ameraucanas. They continued to lay right on through the winter. I didn't get as many eggs, less than a dozen per week, but I still had eggs. I didn't use a single one of the thirteen dozen eggs I'd frozen in preparation for an eggless winter.

I learned an important lesson about food self-sufficiency that first winter. I learned that in the end, we use what we've got. If we have a lot of eggs then we eat a lot of eggs. If we don't, then we eat something else. When eggs are plentiful we eat eggs daily: scrambled, hard-boiled, soft-boiled, over easy, sunny-side up, poached, in quiches or sandwiches, and in custards, puddings, popovers, cream puffs, ice cream, and home-baked bread. Even the cats, dogs, and pigs get a daily egg. I give the extras to family and friends, preserve some for lean egg-laying times, and occasionally trade or sell them. For those lean times I've learned other ways to leaven baked goods, such as baking soda and whey or buttermilk, and sourdough.

My second surprise came that second summer. My chicken research had raised some concerns, because I read that egg production is expected to drop drastically after the first year. So much so, that most producers replace their entire laying stock every year. I wasn't sure what to do about that. For one thing, we wanted to raise our own chicks as replacements, and we didn't like the idea of incubators. We wanted our chicks to be mother-raised. Because of that, I put my hopes on at least one of my hens going broody that second summer, so we could raise at least a few replacements for the following year.

Well, my chickens resumed egg production that second year equal to their first. I know this to be fact because in the early days I kept daily egg counts. I was very happy about this, but puzzled. Why were my chickens so non-compliant in these areas, according to the books?

Now, I'm not one to argue with the experts. True experts deserve a lot of respect. However, was it even remotely possible, that the experts were wrong? I don't necessarily mean in their area of expertise, which is production and profit. To accomplish these goals, "improvements" are necessary: of breed, feed, and management techniques. These include unnatural lighting during the winter and starting each year with a fresh batch of layers, specifically a hybrid breed developed for high egg production. Let's face it, it's more predictable this way for folks in the big time egg business.

Homesteaders and small-scale farmers, however, often choose heritage breeds. We do this not only because the birds are beautiful, but to help preserve rare breeds the world might otherwise lose. Could that possibly be the difference? Could it be that these heritage breeds don't shoot their entire egg wad the first year of their lives, but hybrid chickens do? Is it possible that the old breeds of chickens spread their egg laying out over more years? Or was it something else?

I decided to conduct a little unofficial research on my blog. I asked readers about their chickens: what breeds they had, whether hybrid or heritage, first- and second-year egg production for each breed, and whether they artificially lit their coops during winter's short days.

I had about twenty-five responses to my query and the results were interesting. While the types and breeds of chickens varied widely, almost everyone reported that egg laying remained strong and steady even with three- and four-year-old-hens. In addition, respondents reported decreased egg production at various times, but this was usually associated with molting rather than the number of daylight hours. Most folks responding to my survey did not offer artificial light during winter months. The few who did were in colder climes, where the purpose was for warmth.[1]

The key in this limited sampling seemed to be lifestyle and diet rather than breed. A large percentage of participants mentioned that they allowed their birds to free-range. This not only translates into a more natural diet, but also less stress because the chickens have plenty of room and are allowed to be chickens and do what chickens do.

All of this meant that my original idea to hen-raise several replacement chicks every year would work well for us. My hens would not all be the same age, which would help even out egg production. In addition, we would always have a few elderly hens and young cockerels (roosters under one year of age) which would mean meat for the table. The only thing I would have to figure out is how to discern the age of a chicken.

Of our original mixed flock, only the Ameraucana comes in many colors and markings. Chickens of the other breeds look identical except for sex. This is helpful if one wishes to use some of them for meat. Distinct color and markings seem to lend themselves to identifying, possibly naming, and anthropomorphizing individual chickens. When they all look the same, they seem less pet-like and it's harder to become emotionally involved with them. When it comes time to thin the flock, thoughts don't roam to, "Oh no, how can we get rid of So-And-So, she's my favorite?" They are just chickens.

On the other hand, when it's time to thin the flock, having them all look the same makes it harder to remember which are the ones that need to be culled. I've mentioned replacing the oldest hens, but sometimes an individual is a trouble maker. Such a chicken is forever causing a ruckus and keeping things stirred up. Some chickens are truly mean when they rise to the top of the pecking order, forever pecking,

Free-range chickens are happy chickens. They eat bugs, grass, and seeds.

bullying, and bloodying those lower in the order. At its extreme, this can result in cannibalism and is why confined chickens are usually debeaked. Free-range birds are able to run away as needed, but we usually end up eliminating troublemakers to maintain peace in the chicken yard and on the homestead.

Of elderly birds, I am no expert. Various sources say basically the same things:[2]

>legs become scalier looking
>vents (where eggs come out) become dry and tight
>combs and wattles lose color
>abdomen becomes hard and firm
>space between the keel (breast) bone and vent increases

An easier way to tell would be by using leg rings of different colors each year.

One observation I've made is that older animals eventually seem to rise to the top of the pecking order. They also become less productive as they age so that the least productive animals often end up hogging the most resources! Older animals also tend to have more health problems and require more care. Even their natural death requires disposal of the body. Spending more money on less productive animals may or may not be acceptable depending upon one's goals. For a self-sufficient homestead, this can tip the balance because every critter must carry its own weight. Culling animals by selling, trading, or consuming only makes sense.

With all that book knowledge under my belt, I felt ready to order our chicks. It was an exciting time for us, because after all our dreaming and planning, it meant our homestead was finally beginning to take shape.

BABY CHICKS

The post office called at 5:07 a.m. to let us know they had just come in. Within minutes we were on our way to pick up our very first baby chicks. They had been packed up the day they hatched and would be two days old by the time we got them. Our original delivery date was supposed to be the last week of February, but several weeks earlier the hatchery called and asked if they could move up the date. This was due to availability and demand. It meant a flurry of activity on our part, to finish repairs on the old shed-turned-chicken-coop, but we were ready.

As soon as the postal clerk handed us the small, chirping box, we opened it. We were dismayed to find we'd lost one in transit, two seemed possibly injured, and one was quite weak. As soon as we got them home, Dan put these three in a shoe box as a private little brooder. I dipped each chick's beak in water, made sure they all got a drink, and knew where to find food. We kept a close eye on our precious additions to the homestead. By later that day, we were able to transfer two of the weaker chicks back to the rest of our happy, chirpy, new little flock.

I ordered four breeds, six each of Delawares, Barred Hollands, Welsummers, and Ameraucanas. I ordered "straight run," which means that theoretically I should get half pullets (girls) and half cockerels (boys). All were heritage breed chickens, and picking just four was not an easy task. The more I researched, the more I realized that every chicken owner has different preferences. In the end I chose breeds that I thought were handsome to look at and would give us a rainbow of egg colors. Delawares lay brown eggs; Barred Hollands, white; Ameraucanas, blue-green; and Welsummers, spotted chocolate brown. We hoped to develop a preference for one of them and then keep a rooster from that breed to perpetuate our new flock.

Within the first couple of days, I had to replace the brooder box because somehow they managed to dump out all their water during the night and soak the old box—without tipping over the waterer.

I installed them in a new box, covering the litter with paper toweling until they could learn not to eat their litter. They tend to test everything to see if it is edible. One thing I have learned since then is that hen-raised chicks learn what to eat from their mother. Needing to protect them from eating things they aren't supposed to eat is one chick-chore that can be crossed off the list. Paper towels are better than newspaper for this, because they give them a little traction. Newspaper is slicker and can result in leg injuries.

After a few days I removed the paper towels and allowed them to roam on the pine mulch (purchased in the garden department and much less expensive than pine shavings in the pet department). The chicks immediately began to scratch around in it.

When the chicks were six days old I began to have problems with them. It appeared one had a leg injury. Like the one that was dead on arrival, this one was also a Barred Holland.

Chicks can commonly have a leg problem called "Spraddle Leg." This is a condition where they can't quite manage to keep the leg pulled under their bodies and it keeps slipping out (like doing the splits). Our chicks were having a different problem, however. The one leg was weak, not splayed, and they kept falling to that side.

I scoured the internet for causes and cures. I found information for an "orthopedic chair" for chicks with leg problems, but the instructions didn't transfer the "how to" from the writer's head to mine. I managed to give the chick water with an eye dropper and food with a tweezers, but it continued to grow weaker and started to have trouble breathing. I didn't like to see it suffer, so after much emotional turmoil, I made the heart-wrenching decision to put it down.

I knew that as much as I wanted my critters to live, sometimes death is kinder, even if an animal can be "cured." The question

Barred Holland chick (left), Delaware (middle), and Ameraucana (right). In addition to chick starter feed, I fed them alfalfa sprouts, which they loved. The Barred Holland chicks were the smallest and most fragile. One of the rarer heritage breeds, Barred Hollands are white egg layers and excellent foragers.

shouldn't be "Can I cure it?," but "Am I doing this for the animal's sake, or mine?" It was difficult to make the decision to put that chick down, and it was difficult to do it. But the relief that followed, from knowing that I had made the right choice, more than made up for the inner struggle I experienced beforehand. We had already determined that our chickens were not pets, but livestock. We knew that some of them would eventually be butchered. Still, I wasn't mentally prepared for this so soon.

The next morning I heard a tiny (but loud) squawk from the brooder box. I looked to see one of the remaining four Barred Holland chicks being run over by another. Nothing unusual about that, but when it didn't get up, I took a closer look to discover that this one too, had a bum leg. There had been no earlier sign of it, so I didn't know what happened.

This time I called the hatchery and they were very helpful. They credited my account for three chicks and offered explanations and advice for leg problems. Besides slippery surfaces, leg problems in young chicks can also be caused by a vitamin E deficiency. They recommended a vitamin/electrolyte additive to the chicks' drinking water. Unfortunately, this chick didn't make it either.

The last chick died shortly after that, also a Barred Holland and also with the same leg weakness. There was no prolonged suffering nor a need to do something about it. I was thankful for that. Still, the whole experience left us with a bit of an emotional dilemma to resolve.

We always planned to take the best care of our animals that we possibly could. We would treat their injuries and ailments, but didn't want any of them to suffer needlessly. On an intellectual level, we realized that death is a part of life, and that sometimes life and death decisions have to be made.

This is where the different approaches to keeping livestock part company. The pet-owner or hobbyist will want to make every effort without sparing the expense. Those focusing on production will think in terms of profit and cost effectiveness. For the subsistence farmer or homesteader working toward self-reliance, there is a different balance to consider. Like the hobbyist, we often become attached to our critters; they are a part of our homestead landscape and a part of our lives. But like the food producer, we are aware that any one critter can tip the scales in terms of time and money, so that everything else on the homestead suffers. Although researching our chicks' problems had taken time, it had taken very little money to treat them. That might have been different if I had opted to take them to a vet. I never questioned my decisions, but they gave me something to think about.

At one week of age our chicks discovered that if they ran, jumped, and flapped their wings all at the same time, something wonderful happened. We figured it was time, and so moved them into an enlarged brooder box in the chicken coop.

Wing feathers on a week-old chick.

They were definitely now starting to look like

My cardboard brooder box in the chicken coop was tall enough to keep out drafts. The brooder lamp hangs from the ceiling and can be raised or lowered depending on the chicks' behavior. Huddling under the light indicates they are too cold. Spreading out to the periphery of the box indicates they are too hot.

miniature chickens. They were growing real feathers and combs. The little black Barred Holland chicks were no longer black, but starting to develop their bars. These were the smallest, but boldest of the lot.

By three-and-a-half weeks all the chicks were feathering out nicely. The yellow chicks were no longer yellow, but white. Some of them started to look like little roosters, with larger combs and more upright tails. Also I noticed some posturing behavior—part of establishing the pecking order? Other than that they all seemed to get along just fine.

The largest ones were some of the Ameraucanas. The smallest ones were the Barred Hollands, along with some of the Delawares. Smallest does not equate to most timid, however, and these were the most inquisitive and boldest.

By now they were no longer alarmed when I tidied up their brooder area. I suspected they were beginning to associate me with food, and food is a good thing, right?

When we had a couple of days in the 60sF (upper teens C), I noticed they were all settled down in a ring against their brooder wall. This indicates they are too warm (i.e. they were getting as far away from the heat lamp as possible), so I started to turn off the brooder lamp on warm days.

Four-week-old chicks peeking out of their brooder box. They were curious, but very cautious, when I expanded their territory.

It wasn't long before they began to practice flying. It wouldn't have surprised me if one day I went to the coop and discovered that at least one had managed to fly over the top of the brooder guard. Fortunately that never happened, or else I would have had one lonely, distressed chick.

It was time to enlarge the chicks' territory again. It was mid-March and still chilly and windy enough that I wanted to give them the protection of the cardboard brooder guard and heat lamp. However, they were getting big enough to need a little more room. For their first outing, I opened up the brooder guard to let them explore the coop. We wondered which would be bravest and go first. We thought it would probably be the Barred Hollands, who, though smallest, were the most curious. But it wasn't, it was the Delawares.

Cautious at first, they eventually began to explore their new territory. They went back to the safety and warmth of the brooder at dusk, but a new routine was set for daytime chicken explorations.

In late March it was finally warm enough to let the young chickens out of doors. On the first sunny warm day, we opened the door to the chicken coop to let them out.

At six weeks old, they needed the warmth from their brooder lamp only at night. During the day, they practiced roosting. Even though this behavior comes instinctively, they had to learn how to keep their balance while negotiating the roosting perches.

The Welsummers were the only breed of my four in which I was beginning to see a difference between cockerels and pullets because of the head coloring. They were also the first breed to crow. It wasn't a big rooster crow, but it was definitely a crow.

I noticed one of the chickens taking a dust bath in a patch of damp soil in the yard, and thought it might be a good idea to make them a dust bath box in the coop. I did the requisite research online, and found that folks use various mixes of dirt, sand, dust, hardwood ash, and diatomaceous earth. I decided to use what I had on hand, which was sandy dirt and hardwood ash from the wood stove.

It took a couple of days to dry out the soil, and in the meantime, I found an appropriately chicken-sized cardboard box and cleaned out the wood stove.

Once prepared, I placed the dust bath box in the coop and waited to see what the chickens would do. To my dismay, they were more interested in pecking at the sand in the box than they were in bathing in it. Hmm. Wasn't all that grit I gave them good enough?

Eventually I found one chicken dust bathing in the box, but the real dust bath action turned out to be in the chicken litter on the coop floor. I reckoned I didn't need the dust bath box after all. Live and learn.

By the time the chicks were three months old, we were letting them out to free-range. I still had a couple more months before I could expect my first eggs, and we were still trying to distinguish the cockerels from the pullets.

Six-week-old Delawares and Barred Hollands. They were now being let out into the chicken yard. The feeder was made from four-inch PVC pipe.

By three months of age they looked like adult chickens.

The Welsummers were easiest to sex, because males and females of the breed have different coloring. I had six of this breed, three of each. Of my five Ameraucanas, I'd seen three crowing. Individual birds were easy to tell apart because of the varied coloring and markings within the breed. Of my two Barred Hollands, I seemed to have one of each, the little pullet being darker than her male counterpart. The Delawares were the most difficult to sex. I was going to have to rely on comb, wattle, and tail development after they had matured a little more.

We knew that soon we'd have to start culling the roosters. There were daily mock stand-offs amongst them, but I realized it was going to get more serious pretty soon. At that point I wasn't entirely certain how to choose a flock rooster, except to eliminate some by breed. The original plan was to choose only one breed, and which breed was something we discussed often. Dan liked the Barred Hollands and I liked the Welsummers. We talked about the possibility of keeping two separate flocks, but that would mean creating and maintaining a second chicken area if we wanted to keep the breeds pure. If all we were doing was raising chickens, that would have been a possibility, but considering everything else we wanted to do, it made more sense to stick with one breed. Since Welsummers were more common and easier to come by, the breed of choice became the Barred Hollands. We had a pair and since they laid different colored eggs from the other breeds, I could save out only white Barred Holland eggs to give a broody hen as soon as one became available.

By four months the young roos had begun to mature, and the chicken yard deteriorated from a pleasant, peaceful place on the homestead, to a battleground of ongoing chaos. There were constant squabbles and crowing competitions. The cockerels were continually chasing one another and chasing the pullets. I was beginning to worry that our girls might never start laying because of the uproar. What the books had said was adequate room for nineteen chickens in the coop quickly became too small.

It was at this point that the reality of choosing to eat meat truly hit home. Our plan from the beginning had been to keep one rooster and eat the extras. We understood that we had made this decision on a purely rational basis and that there would eventually be accompanying emotions to deal with as well. But were we ready?

In some ways it seems odd that this should be a question at all. Backyard flocks of chickens were once common in our culture, along with backyard gardens, even in urban areas. Dispatching and butchering chickens were once common culinary skills, practiced by many a housewife. Yet today, the whole process is extremely foreign and distasteful.

There are several options here. Many opt to raise their chickens and then send them to a meat processor to turn them into packages of freezer-ready meat. In some areas it is sometimes possible to hire a mobile processing team, especially for larger animals. For a fee these will come to one's place to do the killing and dressing on site. Some will even haul off the carcass to the butcher for cutting up and packaging. Another option might be to find someone experienced in the process and invite them to teach the first-timer. We had no such options, so we turned to books and videos.

One of the first things I learned was that the term "butchering" does not mean killing. Butchering is the cutting up of the carcass into cuts of meat. The killing part is just that—killing. Politely referred to as "dispatching," it is nonetheless the taking of the animal's life. Processing refers to the entire process, i.e. both killing and butchering.

We also learned that there is more than one way to kill a chicken. The classic chopping off of its head is not the preferred method. Old-timers used to grab the chicken by the head and give it a flip, thereby breaking its neck. This is considered the quickest method, but neither Dan nor I felt comfortable with it. More commonly, a "killing cone"—which looks like an upside down traffic cone with its point cut off—is employed. One places the chicken in the cone, head down, and then slits its throat. To do this humanely, the knife must be very sharp. A quick cut and the chicken bleeds out. There is some flapping and jerking, but the deed is soon done.

Dan and I have worked out a method where I go fetch a chicken and put it in the cone. We start before daylight because it is much easier to pluck a chicken off its roost than to chase it around the yard. He kills it while I go get the next one. When I return, he hands me the bled-out chicken and I take it to our work table before going to fetch another. Once the killing is done we start dunking and plucking. The dunking water should be 150 to 160°F (66 to 71°C). A younger bird needs to be held under water for about 30 seconds (a little longer for older birds). This does make the plucking go a bit easier, although wet feathers do tend to stick to everything.

Once the first chicken is plucked, Dan begins eviscerating (removing the innards) while I continue plucking. We're usually done as dawn begins to break.

We used to divide the chickens into various cuts, but now I simply wrap and freeze the entire bird. It's faster that way and I can cut it up later. If I'm going to can the meat, I cut up first, of course, to fit into the canning jars. I can it with the bones, which are easily removed when the jar is opened to use. Canning is especially good for old hens because the pressure processing helps tenderize the meat.

As I pause in my writing to re-read the foregoing paragraphs, I understand how impersonal and emotionally sterile they seem. The fact of the matter is that the experience of killing a chicken (any animal, really) is an emotionally-charged one. No matter how straightforward the description makes the process sound, the truth is that it is something we dislike doing. We are always concerned about doing the killing as quickly and painlessly for the animal as possible. I confess that we have not always been successful at that, especially in the beginning. The bottom line, however, is that we dislike taking an animal's life. So why do it, you may ask?

This is where the reality of hard choices comes in. We can choose not to kill, but we must still decide what to do with all those rabble-rousing birds. Left to themselves, the roosters will harass the hens, fight, and try to kill one another. Unfortunately, we have had roosters kill each other, and I have to say that fighting to the death is not a kind way to let a bird die. If they don't kill one another off, overcrowding soon sets in. Disease will eventually follow, which is one way of letting nature control the population for us. However, I think vegetarians and meat-eaters alike will agree that this is not a desirable situation. We could try to sell or give away the extras. I have tried to do that and found that roosters are very difficult to get rid of, usually because everyone else has more than enough of their own. One way around killing would have been to order only pullets from a hatchery or breeder. Unfortunately, this is not a true workaround, because

Five of our six Delawares turned out to be roosters.

someone, somewhere, will end up having to do something with those unwanted male birds. On a more far-fetched plane, perhaps someone will develop genetically modified hens which produce female eggs 99% of the time. That may sound absurd, but there are folks out there who actually think this way, who want to re-create nature according to their own view of how it should be.

Taking personal responsibility for eating meat is part of our stewardship. By doing the killing ourselves, we have the assurance that it is done as quickly and humanely as possible. We know that our critters have had good lives, were well fed, and well cared for. Our animals have come to trust us, and because of that, we believe that they do not experience the fear of a potentially terrifying situation. Those cockerels were our introduction to this part of homesteading. In the end, it had to be a rational decision. Once that decision was made, we just did it. We put our emotions on hold and simply went through the steps we had discussed until we were done.

As difficult as the experience was, it was a relief to have all the squabbling and noise replaced with peace and quiet. It was nice not to have greedy bullies in the chicken yard chasing all the others away from anything edible. It was nice not to have to buy so much feed. We started with twenty-four chicks and ended up with one rooster and seven hens.

Our baby chicks were all grown up. Soon they would be on to the chicken business of laying eggs. That was something to look forward to.

EGG!

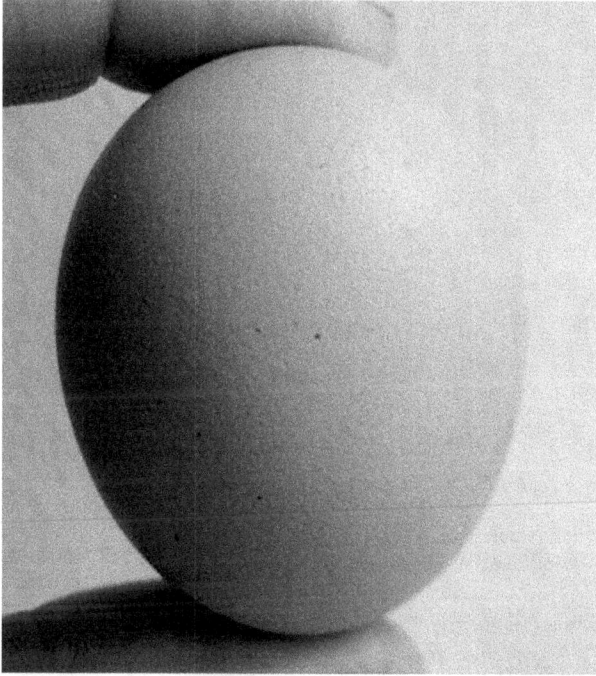

Nothing fills a new chicken keeper with more excitement than the anticipation of that very first egg! Young chickens typically begin laying eggs when they are five months old. Long before ours approached that magical age, we were busy making preparations.

Preparations meant nest boxes. We could have bought them ready-made, but we decided to make ours. We took a bit of time to research this and discovered that people made them out of wood, sheet metal, even 5-gallon plastic buckets or totes. We decided to use materials we already had on hand.

We made ours from huge cardboard tubes Dan found in a Dumpster. He cut them and mounted them to a couple of two-by-fours, and I painted them. I called them our "3R" nest boxes, based on the recyclers' mantra of "Reduce, Reuse, Recycle." I was exceedingly pleased. We installed them in the chicken coop when our pullets were about eighteen weeks old. I placed a couple of golf balls in one nest to give the hens a hint, and then we waited.

Our newly installed 3R nest boxes with golf balls to give the hens a hint.

The first egg came from our Delaware hen when she was nineteen weeks old. That's earlier than the books say, so that was exciting too. Did she lay it in one of my 3R nests? No, she laid it underneath.

I didn't realize it at the time, but this was my first lesson that chickens have minds of their own. We might think of them as "dumb clucks," and they certainly wouldn't make very smart humans, but neither would humans make very smart chickens.

Once all seven of my pullets were laying, I was getting five or six eggs per day. Nary a one was laid in those nest boxes. They laid them under or next to the boxes. The Ameraucanas turned out to be flyers. These would use their wings to fence-hop and lay their eggs in the carport where we stored the hay.

This is a problem with free-range chickens—they tend to lay where they want, rather than where you want. On top of that, they will periodically change that laying spot, usually when you've figured out where it is.

Of homegrown eggs, I have to say that nothing beats them. Their rich golden-orange yolks are glorious to behold and give lovely color to scrambled eggs and baked goods. Everyone I've shared eggs with tells me how much better they taste than store-bought eggs. One neighbor was worried when he went to the doctor, because he feared his cholesterol would be high after eating so many of our eggs. To his

Hens are called pullets during their first year of life, and their first eggs are typically small. Our first egg was 1.1 ounces and about the size of a golf ball.

amazement, his cholesterol was fine. Considering research performed by *Mother Earth News*, this isn't a surprise. In recurring tests of free-range chicken eggs, results have consistently shown that homegrown, free-range eggs contain 1/3 less cholesterol, 1/4 less saturated fat, 2/3 more vitamin A, two times more omega-3 fatty acids, three times more vitamin E, seven times more beta carotene, and four to six times as much vitamin D than eggs from confined, factory chickens.[1, 2, 3]

For us that's certainly good news, but the bottom line is that chickens—and eggs—are a vital part of our homestead.

THE CHICKEN WHO THOUGHT HE WAS A GOAT

Every time I think I've got my chickens figured out, they do something new to amaze me. I've learned a lot from observing them, and also from reading books and the internet. But here's something I hadn't read about anywhere and to be honest, I'm not sure it's "normal."

Dan and I had agreed we wouldn't name our chickens since they weren't pets, but even so, I began to call our chosen rooster Lord Barred Holland after his breed. Lord B loved the goats. He especially loved CryBaby. When I let the chickens out to free-range in the morning, he immediately ran off to find the goats. When he found CryBaby, he would run up to her and do a little happy dance. Then he would follow her and her mother around, accompanying them for quite a bit of the day. Wherever you saw the goats, you didn't have to look far to see His Lordship there as well.

He didn't completely ignore the chickens; it was just that he seemed more interested in goat business than chicken business. To his credit, he always came to greet me when he saw me, but then it was off with the goats again.

One day, while standing in the checkout line at the farm supply store, my eye happened upon the latest issue of *Barnyard Gossip* magazine. I couldn't resist a sneak peek at the cover story!

.... read it here first

Barnyard Gossip

Breakup!

Is the great romance over?

Was it doomed
from the start?

hunk roosters

lgd buddies

cute kids

tales from the coop

fashion halters

grain taste test

ask Nanny

and more!

who is

Crybaby

seeing now???

July 2010
priceless

0 71361 06121 7   01>

Insider sources tell us that the great barnyard romance between
Lord Barred Holland and CryBaby appears to have come to an
end.

"I just woke up one morning," stated Lord B, "and realized,
'Hey! I'm a chicken!'"

His Lordship has recently been seen in the company of a
bevy of beautiful pullets, strutting his stuff and chasing away the
other cockerels.

CryBaby was not available for comment, but according to
her mother, she responded to the rumors by saying,

"Lord Who?"

Exclusive shots taken by our staff photographer have confirmed speculations that she is indeed seeing someone else.

While his identity remains a mystery, Barnyard Gossip has verified that he is an Ameraucana cockerel. Gossip mongers report that he can be seen on the coop roof first thing every morning, after being chased around the chicken yard by Lord B himself. Mr. Ameraucana walks along the roof line and flies down into the goat field, where the chickens currently free-range in the afternoon. The chickens aren't allowed into the field until later, so he has the goats to himself all morning.

Of the Ameraucana's daily escape, coop keeper Leigh stated, "It's just as well. Otherwise he runs into the coop and tries to hide in Mrs. Delaware's egg-laying spot. Not that it does him any good, because his tail sticks out and he talks to himself the entire time he thinks he's hiding."

Of Lord B's sudden interest in the hens, she said, "I think it was basically a maturity issue. According to The Livestock Conservancy, Hollands have a slow-to-moderate growth rate, meaning they are slower to mature than many breeds. Once Lord B's hormones kicked in, he knew where he belonged."

She also noted that Lord Barred Holland still does his happy dance when he greets the goats. "He just doesn't hang out with them anymore," she said.

When asked about his rival, Lord B was overheard to say, "He's soup."

You read it here first, folks.

ODE TO LORD "B"

A place of honor is set aside,
For one whom respect is due.
Always watchful, always near,
Everyone concerned in view.
And when a squawk comes from afar,
He's up and on his way,
He does not dawdle, he is not slack,
There is no slight delay.
He walks about, his head held high,
His voice for all to hear.
To let you know that all is well,
If not then give an ear.
The day has dawned,
A crow rings out,
Wake up, wake up, tis morn.
The sun's anew,
There's work to do,
So get your coveralls on.
There are goats to feed,

There are stalls to clean
There are sheep that need a shorn.
I'm just the one to tell you so
Beware and be forewarned.
I'm in charge of this here farm,
Don't lie or muck about.
Come do your chores
And don't be slack,
Or I will roust you out.
I proved myself among the rest,
Foraging and guarding hens.
Oh yes, it's obvious to all,
That I hands down am best.
A Barred Holland rooster is what I am
But just call me Lord "B."
Lord of the land and my domain
Why don't you come and see.

By Daniel Tate

BROODY

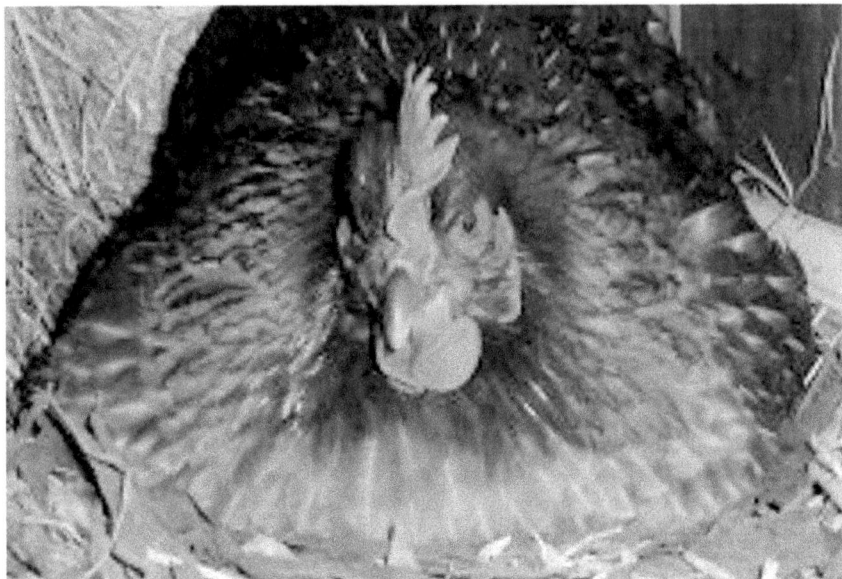

One of our self-sufficiency goals is to perpetuate our flock of chickens with hen-raised chicks. Anyone with chickens can tell you this is rather iffy, because it requires a cooperative hen that is willing to set on a clutch of eggs for twenty-one days and hatch them out. Some breeds are more prone to do this than others, as are some individuals, but even then there are no guarantees. It's why many folks prefer to use an incubator. However, we determined in the beginning that our animals should be allowed to live their lives as naturally as possible. For chickens, that meant hen-raised chicks.

One day I went out to check for eggs and discovered that the most favored egg laying spot (the one beneath the nest boxes) was occupied by one of my Welsummer hens. I didn't think much of it except that it was odd that she made a soft, high-pitched trill any time another of the hens came near. I'd never heard our chickens make that sound before. Even odder, Lady Delaware, who was at the top of the pecking order (and very demanding when it came to her right to lay eggs wherever she wanted), respected her for it. That afternoon the makeshift nest was empty. I collected the egg from it and assumed it was chicken business as usual.

After sunset, I made my final chicken count of the day before closing the coop for the night. I counted only seven chickens on the roost. It should have been eight. Puzzled, I wondered if something had happened to one of the hens. I hadn't heard hawks that day, but that didn't mean they weren't around. On a hunch, I peered into the dark, to see if that nesting spot was occupied. Sure enough, Mrs. Welsummer was in it.

The next morning she was still there. I thought this was unusual because Welsummers supposedly go broody only rarely, so she was the last one I expected to do so! In spite of that, this Dutch breed is popular with homesteaders not only because they are dual-purpose and excellent foragers, but, especially, because they lay dark brown spotted eggs.[1]

I read that to have the best hatch rate, the eggs given to a broody hen (or placed in an incubator) should not have been refrigerated.[2] Happily, I had a dozen eggs sitting on the kitchen counter from the previous days' collections. Mrs. Welsummer let me poke eight of them under her before she gave me a peck. Later, when she got up for a brief break, I counted nine eggs in her nest. Someone had laid another egg.

The eggs were a variety from our four different breeds of hens. If my broody Welsummer maintained the twenty-one day vigil and they did hatch, it would mean a mixed breed batch of chickens. That was something of a departure from my plan to raise Barred Hollands. Originally, I thought I'd keep Barred Holland eggs at the ready, in case any hen went broody. With only one Barred Holland hen for my Barred Holland rooster, however, that was easier said than done.

I suppose I could have waited and collected a batch of white Barred Holland eggs. In a natural setting a hen will lay for several days and then commence setting. When the eggs start to hatch, she will wait several days before getting up, to allow all those eggs to hatch. In my excitement over our first broody, however, I just used the eggs that I had available!

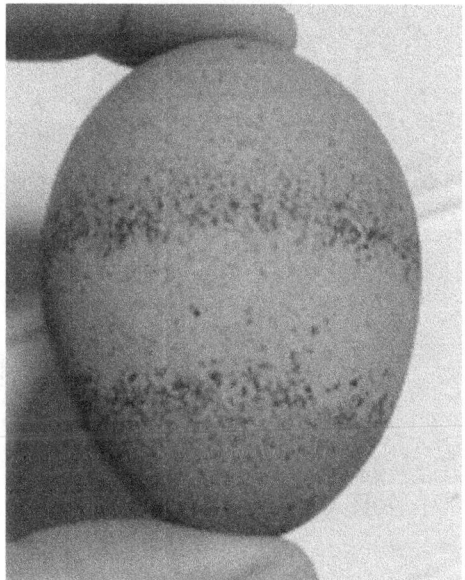

Welsummers lay interesting eggs.

We did have some question about how many of those eggs were fertile, or if any of them were. We'd watched Lord Barred Holland with his ladies and it seemed to us that his, um, aim was off. I read up on candling eggs. Candling is a technique which involves shining a bright light through an egg in a dark room. The light somewhat illuminates the contents of the egg. An infertile egg will appear basically clear; a fertile egg will contain a dark blob (the chick). This is easier to do with white or light-colored eggs than dark ones.[3] I figured that if none of the eggs were fertile, I'd institute Plan B. Plan B would be purchasing chicks and popping them under Mrs. Wellie at night on day twenty-one. I'd heard mama hens will raise grafted-on chicks as though they hatched them themselves. My eggs had mostly dark shells and were difficult to candle. When I couldn't tell what was what, I went ahead and ordered chicks from a hatchery. We had wanted to try Buff Orpingtons so I figured this was my chance.

A few more observations about broodiness: a broody hen is tenacious! If you toss her out of the nest she won't land on her feet;

This is the cold water dunk method of trying to break a broody. I submerged her up to her neck for a minute or two. Unfortunately, this method did not work for me. However, it taught me the meaning of "madder than a wet hen."

she'll just topple over in her puffed, ruffled position as though she isn't thinking about anything else. And she isn't. A broody hen is about one thing, and one thing only: hatching eggs and raising chicks.

Broodiness is caused by the release of the hormone prolactin.[4] Besides stopping egg laying, the broody hen's body temperature rises. If a broody hen isn't wanted, the logical conclusion about breaking broodiness is to lower her body temperature. Some say they've had success by removing a broody from the nest and locking her out of the hen house. Others use a wire-bottom cage to cool off her bottom. When nothing else worked I once tried the cold water dunk method.

Unfortunately this didn't work either. However, according to the Wikipedia article cited above, neither do hormone injections. The one method I have yet to try is ice cubes. It is said that placing the ice inside plastic Easter eggs and putting them under the hen will work. If I ever do try it, I'll report the results on my blog.[5]

For me, broodiness hasn't been a problem; but then, I've only had one or two hens go broody at a time. If I had three or five or six, I'd definitely take measures to do something about it because that would mean a drastic decrease in our egg production. Otherwise I only give a broody hen as many eggs as I want chicks plus a few extras in case some of them don't hatch. That way we get to enjoy another blessed chicken event, and my broody hen is happy.

Mrs. Welsummer hatched out two of the six eggs I gave her, including one, hoped-for, Barred Holland. She readily accepted the sixteen mail order chicks, and it was a happy time on the homestead.

MRS. MEAN

I have already mentioned that I usually don't name my chickens. On occasion, a few have had such outstanding characteristics as to earn themselves a nickname, but for the most part, I call everybody "Mrs. Chicken." Except, of course, our rooster.

One of my Barred Hollands earned herself a place on the name-exception list. She became known as "Mrs. Mean." In our original flock, she was at the very bottom of the pecking order. She was constantly being picked on, pecked on, and chased away by the others. She had so many feathers pulled out that she was forever half-naked. When our first batch of homestead baby chicks hatched, she made every effort to ensure a new position for herself in the pecking order. Under no circumstances was she going to allow the newcomers to rise above *her*.

Barred Hollands are an American heritage breed of chicken. They were developed in the 1930s and are one of the few dual-purpose chickens to lay white eggs. The dual-purpose breeds were preferred by small farmers due to their versatility. When white eggs became the consumer rage, the breed was developed by Rutgers Breeding Farms. Winter hardy, they are excellent foragers and find quite a bit of their own food. If I had occasion to use a shovel in their free-range area, I could expect the Barred Hollands to be right there in my business, looking for grubs, seeds, and worms. Eventually the breed lost popularity and its current population is listed as "critical" by The Livestock Conservancy.[1] While many homesteaders prefer brown egg layers, the Barred Holland is nonetheless an excellent choice for homesteaders.

The temperament of this breed is characterized as calm. From personal experience I have to add that of the various breeds we've tried, our Barred Hollands definitely had the most personality. For example, the very chicken we're talking about. While very sweet and friendly toward humans, Mrs. Mean was definitely not sweet toward Mrs. Welsummer's brood chicks. Most of the time she chased away whichever one was handy. Or gave them a peck, or yanked out a feather just to remind them who's who. On occasion though, she got her ornery on. One day I heard panicked squawking and a frantic flurry of wings in the hen house. Fearing some wild critter was after the chickens, I took off running. Turned out Mrs. Mean had decided to occupy and defend all three of the nest boxes and was chasing all the pullets out of the hen house. I removed her to the goat shed and egg laying resumed peacefully.

Another time, I had a Buff Orpington that went broody and was trying to set on a clutch of eggs. Mrs. Mean kept chasing her off the nest. She would jump into the broody's nest and stand on her head until Mrs. Broody finally had to leave. Mrs. Broody would situate herself in another nest box, but thanks to Mrs. Mean, the eggs got left behind. I would move the eggs to Mrs. Broody's new nest box, but Mrs. Mean would chase her off again.

As you can imagine, this game of musical nests quickly became a nuisance. Broody hens, however, can be extremely persistent. Mrs. Broody finally settled for a spot under the nesting boxes, where Mrs. Mean could no longer jump on top of her and stand on her head. I let Mrs. Broody have three eggs. She hatched out one, but that's another tale to tell.

Mrs. Mean did not have a happy ending. We had discussed eliminating her, but never got around to it. Then one Sunday, Dan discovered our Great Pyrenees puppy running around with a dead, mangled chicken in his mouth—one of our two Barred Holland hens. Since those hens were identical I was never sure which one he got. Either the pup got Mrs. Mean, or Mrs. Mean was smart enough to behave herself after that.

CHICKEN LITTLE

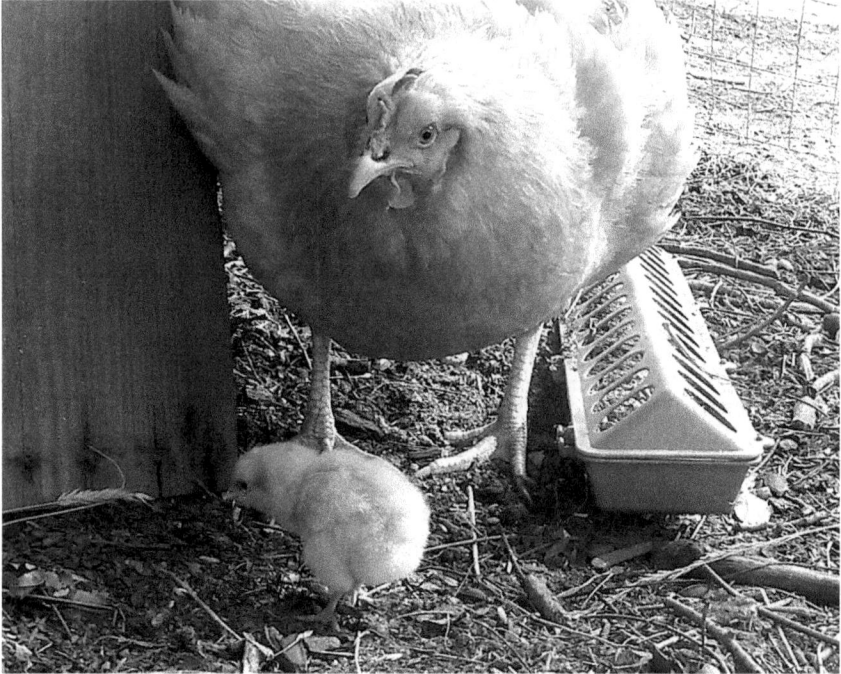

This is the story of Mrs. Broody and Chicken Little. Before Chicken Little was hatched, his mother, Mrs. Broody, was very persistent in wanting to hatch some eggs. At the time I wasn't ready for a batch of baby chicks, so I attempted to break her broody with the cold-water-submersion method. This lowered neither her body temperature, nor her desire for a clutch of eggs, but neither did Mrs. Mean standing on her head. I finally gave Mrs. Broody three eggs, jotted down the date, and forgot about it.

One day during evening milking I heard this mama-wannabe clucking softly. I stopped to take a closer listen and heard a tiny "peep, peep, peep." If I tried to come near, she would clam up and hunker down. The peeping would stop too. Since my Buffs were mama-hen raised, none of them are particularly friendly, so I backed off and waited. Eventually I had a chance to see the new chick, another Buff Orpington. Of the three eggs, it was the only one that hatched.

Because of previous problems with chicks being rejected by the flock, I put Mama and Chicken Little in a pet carrier in the chicken yard. They needed to be protected from predators and overly playful puppy dogs as well. As pleased as I was with my makeshift home for them, Mama was less than impressed and moved herself and Chicken Little into the hay feeder in the goat shed.

Acceptance into the flock was smoother than anticipated, with Mama getting picked on the most. I think these challenges were from hens lower in the pecking order, hoping to gain a higher position. There was only one challenge each, and the flock order resumed.

Eventually Mama Hen decided it was time to start spending the night in the coop. She set about trying to coax Chicken Little to follow her up the ramp and into the coop. Doing this, however, was a little trickier than simply walking through a little door. Because of the way the building was constructed ninety years ago (with diagonal bracing in the corners), we built ramps to give chickens access to the outside.

Was Chicken Little reluctant because the ramp was so steep?

Was it because it was so dark and scary-looking inside the coop?

Was it because there were big mean chickens inside?

Mama Hen would go into the coop, to the feeder, and cluck loudly in her best "I found food!" cluck. Chicken Little would "peep peep" like the dickens outside, but would not venture into the coop. This went on for a number of days until finally Mama started spending the night inside the coop anyway. Left to his own resources, Chicken Little continued to live with the goats.

Well, I thought, *this will never do*. I didn't want Chicken Little to spend the night alone, so for several nights I would go catch him and carry him into the chicken coop.

One evening, about four or five days later, I went into the goat shed, but there was no Chicken Little. I listened, but didn't hear any peeping. I looked around, but saw no chicken anywhere. I finally took a look in the coop to discover that Chicken Little had figured it out and gone to roost without help.

He was one of the big chickens at last.

Chicken Little eventually grew up into a real rabble-rouser. Once he matured, there was constant chaos in the chicken yard, indeed, on the entire homestead. He was forever chasing the hens, who were forever running away. When Cowboy (the resident rooster at the time) wasn't chasing him, there were ongoing crowing wars. Chicken Little would ambush the hens—even in the hen house—so that they wouldn't go in to lay their eggs. One interesting observation was that the hen being chased would always run to Cowboy. When she got within a foot of him she would stop and look at Chicken Little as if to say, "Safe!"

One day when the ruckus was particularly bad I asked Dan if he'd like rooster for dinner. We did the deed and peace and calm immediately descended upon the entire homestead. It was a tremendous relief. I'd like to think we gave Chicken Little a good life, but he and his own kind made that nearly impossible.

THE SOCIAL INTEGRATION OF CHICKENS

The previous tales have given you a glimpse into the challenges of adding new chickens to one's flock. If you have chickens and have ever tried to add more birds, you are likely nodding your head in appreciation of the difficulty of the task. I don't have specific solutions, and everyone's story will be different. What I can share is what we've experienced, what I've tried, and the lessons I've learned in the process.

The production approach to keeping chickens is to replace one's entire flock every year, so that social integration isn't an issue. As a homesteader with a sustainability approach, my goal is to have a few new chicks every year to replace the oldest hens in the flock. I'm not necessarily interested in increase, I just want to maintain enough chickens to meet our needs with no more than a few extras. One of my concerns has been how well new chicks will be accepted by our existing flock. I did a lot of reading about other folks' experiences with this, and

The concern with new chicks is how well they will be accepted by the flock. The mama hen often has to fight to regain her place in the pecking order as well.

with great interest. I read reports that spanned the gamut; from roosters helping feed the new chicks to roosters killing them and/or the mama hen.

Partly because of the set-up in our coop, and partly because I was unsure of how the older chickens would react, I set up a hatching and baby chick area. I used one of my two goat stalls, separated from the rest of the barnyard by a welded wire fence. In addition, I fenced off an outdoor chick yard where the mama could take her chicks outdoors undisturbed.

Mrs. Welsummer and her brood were the first occupants of this arrangement. The chickens could see and smell the chicks from the beginning, but could not interact. They would stand around and watch them through the fencing, as though curious about their existence. Eventually, the chicks started slipping out through the fence and under the gates, and a few of the hens started entering the chicks' yard. Some of the hens would chase the chicks, and in turn would get chased by Mama Welsummer. Gradually they began to mingle somewhat, although Mama was always wary.

I knew that part of the chicks' social integration would be establishing their place in the pecking order. Anyone who has observed animals for a length of time knows that their social structures are vastly different from that of humans. Humans make a political pretense of trying to establish fairness and equality. Animals, on the other hand, don't give a flying flip about that stuff. They all know that somebody is at the top, somebody is at the bottom, and everyone else is in between. Unlike humans, however, every animal knows its place and accepts it. Occasionally one will challenge another, but it is soon settled and everyone accepts the outcome. No one's feelings are hurt or self-esteem damaged because of their rank in their species' social order.

Eventually, the chicks began to mingle peacefully with the older hens, although there were two chickens which continued to be aggressive toward the chicks. These were the Barred Hollands, Lord and Lady B (aka Mrs. Mean). I don't know if this is a breed trait or their individual personalities. Mrs. Mean, being at the bottom of the pecking order, seemed determined not to remain there under all the newcomers. I wasn't sure of Lord B's status, but I knew it wasn't at the top. The Delaware hen had the top spot—she, being the only one allowed to peck at him. (Yes, the well-known term "henpecked" really does come from chickens!) No matter, he was extremely possessive of anything he thought his ladies might like to eat, and extremely territorial of both the coop and the chicken yard. Sometimes he would chase the chicks relentlessly; other times he would ignore them. Whenever he let out a loud squawk, the chicks would hightail it for quick cover.

Eventually most of the chicks were able to roost in the coop. By that time the older chickens seemed to accept them pretty well. Lord B, however, would still chase and pin one down occasionally, but I was usually able to distract him. The only time I had hope was when we had a hawk attack. Lord B actually helped round up the chicks and get them under cover. This truce was short lived, however, and when it became apparent that His Lordship would not accept the chicks, neither cockerels nor pullets, we had to eliminate him. We could not keep a rooster who would not cooperate with our chicken management plans.

I took great care in choosing our next rooster from the up-and-coming Buff Orpingtons. Of our fourteen chicks, eight turned out to be cockerels. They were different from our first batch for several reasons: all but one was the same breed (Buff Orpington), they were hen-raised, and there was an adult rooster already on the premises. A rooster that chased them all over the place, I might add. With our first batch of chicks it became every rooster for himself. They competed with one another for the food and the hens. This new batch of cockerels, however, banded together as a pack. They would forage together with one keeping watch and giving warning if I walked around the corner. They would chase, corner, and jump the hens. One would pin her down while the others all pecked at her head.

All of this caused no little upset in my original flock, especially once Lord B was gone. Mama Welsummer had a place in the new pecking order (probably because she had raised them), but the other Wellie and Barred Holland refused to leave the coop. The Ameraucana sisters vacated it altogether, taking up residence in the garden tool shed.

Initially I assumed that these behaviors were happening because Lord B had been so dominant. I assumed once he was gone things would settle down and another rooster would naturally take his place. That wasn't the case. There were changes in the cockerels' social structure as we thinned their numbers, but none became any more attached to the pullets and hens than any other. They remained their own team.

I watched them carefully over many days, trying to figure out which rooster would be best to keep. At first I looked for the characteristics that make for a good flock roo: alert, observant, deferential to the hens, and not too "friendly" toward humans. What we humans assume is friendliness in a young rooster is actually boldness. While it's endearing to have a young rooster come right up to you, this boldness often becomes aggression as he gets older.

As we gradually thinned out the most aggressive cockerels, I realized that flock dynamics remained the same. The next guy in the pecking order simply moved up and nothing really changed. In the end, my choice was not based on personality characteristics, but because I noticed that we had one rooster that didn't hang out with the other Buff Orpingtons as much. More often, he could often be found with my older hens. They in turn, kept a wary eye on him, but didn't run in terror every time they saw him. I even saw him try to sweet-talk one of the older gals one time, offering her some tidbit that he held in his beak. I figured if I kept this guy, he would eventually be accepted by all the hens, and I might have a chance of restoring peace in my chicken yard. In addition, he was one of the largest cockerels, and size is a good quality to pass on to offspring in a dual-purpose flock.

One thing that became immediately obvious was that this rooster was not as romantic with the ladies as Lord B had been. Lord B used to sidle up to his intended and dance when he was in the mood. While she stood there, mesmerized by his fancy footwork, he would hop on and do the deed. This new roo, however, announced his intentions by puffing out his wings and chest, charging his victim like a raging bull, and grabbing on at a gallop like a rodeo rider. Consequently, he earned himself the nickname of "Cowboy."

With only one rooster, there was less crowing and no competition for either food or hens. The chicken yard was quieter and calmer on that score. I watched to see how Cowboy would settle in and how the girls would like him. Initially he spent more time with the older girls, but gradually migrated toward the Buffs, who were his hatch-mates.

Eventually things settled down and all the chickens began to spend the night in the coop once again; even the Ameraucana sisters consented to come back. Even so, it was as though we had two flocks which simply shared the same quarters. Cowboy began to spend most of his time with the other Buffs and would chase the older gals back into the coop if they tried to come out into the yard. Perhaps the only original hen who was happy with the new arrangement was Mrs. Mean. She was no longer at the absolute bottom of the pecking order. Now, she was closer to the top, over all of the Buff pullets. She was a happy camper.

By that time I realized that nothing stops the initiation rites amongst chickens. The pecking order is very specific and no chicken wants to lose her place. The oldest chickens are almost always on the top and therefore have first rights to anything chickens want: food, tidbits, water, and the topmost roosting bars. Eventually, after much

Cowboy on the right. With the change of roosters, Mrs. Mean (left) was the only one of my original chickens to gain a higher status in the pecking order.

pecking, many squabbles, and considerable chasing, every chicken knows his or her place in the chicken order of things. As much as I had wanted to protect the new chicks from this, I realized I couldn't. The best I could hope for was that they reached an understanding quickly and that no chicken was totally rejected or killed in the process.

With that knowledge under my belt, I decided to try something different with our next batch of chicks. Rather than setting up the broody hen in her own area, I decided to fence off a nesting and brooding area within the coop itself. I hoped that this way, the chicks would be seen as part of the coop territory. I kept them in their brooder area until they got too big for it.

This new method seemed to work pretty well. While the hens didn't bother Mama and her brood, the biggest trouble maker was Cowboy. He seemed to take great pleasure in tormenting the growing chicks. He loved to sneak up from behind and grab a beakful of feathers. He would strut around proudly with those feathers in his beak while the youngster ran away squawking. In the end, Cowboy was too much of an alarmist and seemed to keep the entire homestead in an uproar. It was time for a change.

Our next broody hen was given six Buff Orpington eggs to hatch. In addition, we gave her chicks of other breeds we wanted to try—Silver Laced Wyandottes and Speckled Sussex. We really liked the Wyandottes, so Cowboy and the three Sussex cockerels were the first elected for "freezer camp." That left us with about twenty hens of six breeds and three Wyandotte cockerels. Happily the chicken yard battles gradually began to smooth out. Something that seemed to help was moving the feeder out of the hen house and into the chicken yard. Food is always something about which animals are possessive, so I figured moving it outside would make the coop a more neutral territory. I covered the feeder at night and moved it back inside when it rained. This eliminated much of the squabbling that used to take place inside the chicken coop and I began to notice all the chickens simply hanging out together.

We chose our Wyandotte rooster by the process of elimination. I took no special efforts to choose a particular rooster, because they all looked so much alike that I couldn't tell them apart. We had three Wyandotte cockerels and the one who didn't get caught during culling became the new flock rooster. We called him "The Sultan."

I had two hens go broody under The Sultan's reign. Both times I gave each of them four or five eggs in a corner of the chicken coop. I put a hardware cloth guard around the setting hen, so that the others wouldn't try to lay more eggs in her nest. The guard was removed a few days after the chicks hatched. Since the other chickens tend to gobble down whatever I give the mamas and chicks to eat, I closed the coop door several times a day while the others were out in the pasture. Mama and the chicks got clean water and plenty of scratch and chick feed. Then I would reopen the coop door to give the hens access to the nest boxes once again. Amazingly, the chicks were all accepted as part of the flock. Not one chicken, including The Sultan, challenged either mama hen or her chicks. In fact, The Sultan seemed to take no notice of them.

In the end, my worry and interventions regarding integrating new chicks into the flock proved ineffective. Letting the chickens sort it out themselves has proven to be not only more effective, but simpler. That doesn't mean I still don't worry somewhat about little ones getting hurt, but I have learned that in the end, I really can't change anything. I do think the rooster is key, but I have learned that the instinct toward a pecking order within a species can't be reasoned or trained out of it. Kindness, acceptance, and respect are choices of will which only humans can exercise. In the end, it's better to let things take their natural course in my critters' various worlds and accept the results.

MAIL ORDER CHICKS FOR A BROODY MAMA

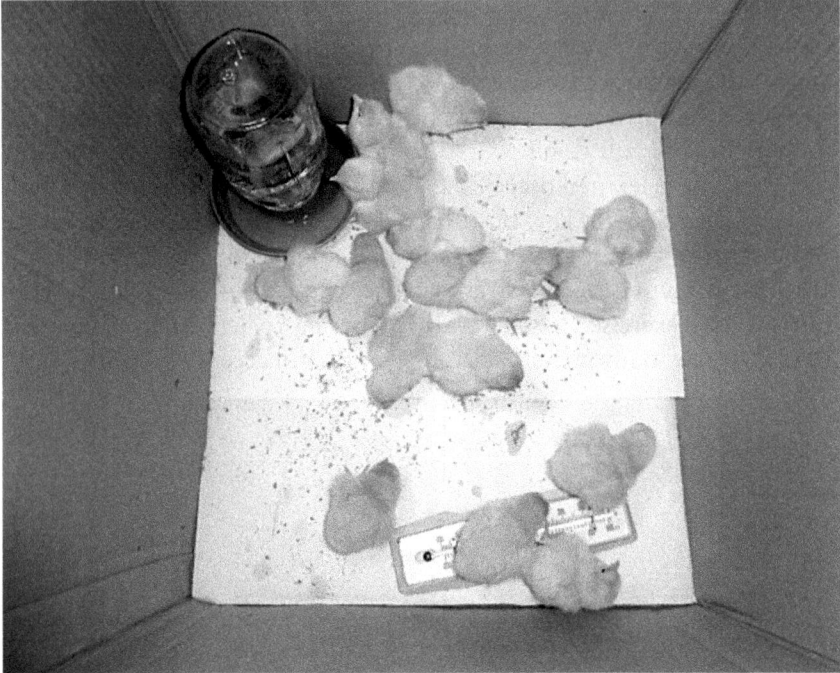

Everything should have a mother. At least, everything God intended to have a mother should have a mother. In spite of the role of a hen in hatching and raising baby chicks, the amazing thing about chicks is their ready-made instincts to find food and water from the get-go. This lends them to being hand-raised rather than hen-raised in most situations. In fact, many farmers and homesteaders prefer to use an incubator for hatching eggs. Dan and I prefer to do things the natural way, although it necessitates having a hen willing to be a mother.

I have certainly found it easier to let a mama hen hatch and raise the chicks. She keeps the eggs at the perfect temperature and humidity and knows when to turn them. No heat lamp and electricity required; Mama squats down, all fluffed up, for any chick that needs to run under to warm up. She takes care of pasty bottom as well, keeping their little bottoms clean. Pasty bottom is the condition whereby chick feces become a little runny and stick to the chicks' bottoms. This usually happens because of stressful conditions such as shipping. If the chicks' bottoms are not kept clean, the feces build up until the chick gets clogged up and dies. With no mama hen, it's up to the chicken keeper to do this. If hen-raised, the mama will do this for you.

The downside to hen-raised chicks is that they don't become socialized toward humans as easily. Their mama runs interference when you're around, always placing herself between you and her chicks. She also places herself between the other chickens and her chicks, or the cat and her chicks; she is simply protecting them.

This may not be desirable for anyone who wants truly "friendly"* birds. Some folks like to be able to pet or pick up their chickens. I have to admit that with our first hand-raised flock, it was a good feeling to have them come running anytime they saw us. Our first rooster, Lord B, would invariably accompany us anytime we took a walk around the property. On the other hand, they were sometimes underfoot. If we were digging holes for fence posts, for example, we would have chickens in the way scratching for worms, grubs, or other buried goodies.

Eventually hen-raised chicks figure out that humans bring food, so they become more tame as they get older. The only exception I've had to this was my first batch of Buff Orpington chicks. At several years old, they still flee whenever they see me. There is a reason for this, however. When they were babies, Mama Welsummer would take them on foraging expeditions. That was fine until they started venturing toward the front of the property. I wanted to keep them out of the garden (chickens can be very destructive in gardens) and away from the road. I would take a big bath towel, walk behind them, and flap it crisply. As they fled in terror I would call out "Shoo! Shoo!" I hoped it sounded like hawks wings and would possibly help keep them safer from hawks. Perhaps it did, but they forever associated those traumatic episodes with me, and have run away from me ever since. Or they at least keep a suspicious eye on me when I'm around.

When we had our very first broody hen, Mrs. Welsummer, I was excited about the possibility of our very first home-hatched chicks. By that time we'd already decided against keeping Barred Hollands and Dan wanted to try Buff Orpingtons. I read about grafting baby chicks

---

*"Friendly" is a human term. It is based on the belief that an animal is friendly because it is nice or because it likes you. While some animals may bond with humans, in general, "tame" is probably a better term. A tame animal is used to human presence and handling. With chickens, particularly roosters, boldness is sometimes confused with friendliness.

So far I've had success in grafting baby chicks onto broody mama hens.

onto a setting mother hen and decided to give it a try. Not everyone has similar success, but since I've not had a problem doing this I'll pass on to you how I did it.

1. Order the chicks to be delivered as close as possible to hatch day.
2. After they arrive, keep mail order chicks in a brooder box with heat lamp until the eggs under the broody hen begin to hatch.
3. At night, gently slip the chicks under mama hen one by one.
4. Worry a little all night about what will happen in the morning.
5. Hopefully success!

The first time I did this I was concerned because the mail order chicks were several days larger than the newly hatched ones. But the littlest knew to find their way farthest under mama while the older chicks huddled closer to the outside, so all was well.

So far every broody I've tried this with has readily accepted all the chicks as her own. We've been able to try new breeds of chickens, and they get the benefit of having a mother. It's better for the chickens, and work-smarter-not-harder for us.

CHICKEN HAWK

I was in the kitchen washing dishes when I heard Dan fire off his .22. I didn't think anything of it because he usually uses it to scare off dogs, which are drawn into our yard like magnets because of the chickens. Unless a dog is bred and trained for farm purposes, even a "good" dog will chase and kill chickens for the sport of it. This is not a big problem for us, but on occasion we have stray dogs passing through.

A few minutes later he came into the kitchen. "Didn't you hear the chickens?"

"No," I replied. "I heard your gun, but I didn't hear any fuss from the chickens."

"A hawk had one of them," he replied.

He had fired a shot to scare it off, and fortunately it flew off leaving the chicken behind. From a distance, Dan couldn't tell which chicken, so I ran out to check. Several of the hens were cowering wide-eyed in the goat shed. Our rooster, Lord B, was outside the shed strutting back and forth and fussing up a storm. The chicks were scattered. A half-dozen were clustered around Mama Hen, two were hunkered down in the field by the fence, and the rest were hiding under the bushes and brush. I called "chick, chick, chick," and when they saw me, they made a mad dash for Mama. I counted sixteen, and then went to look for the other hens. Soon everyone was accounted for, looking scared but unscathed.

Now, before anyone gets up in arms (is that a pun?), yes, we know it's illegal to shoot hawks. But it's not illegal to scare them off. Yelling at them or throwing things doesn't do the trick, trust me. We have to do something because we've had hawks try to nab our chickens previously.

The first attack was on one of our Welsummer hens the year before. I was out turning the compost pile when I heard a crash and Lord B let out a loud warning squawk. I looked up, expecting to see that a pecan branch had fallen down near the chickens. What I saw was a hawk, extracting itself from the thicket under the pecan tree where the chickens like to hang out. Fortunately, its talons were chickenless, but I had some panicked pullets who didn't know which way to run. I ran over to see if everyone was okay. The near victim looked like she'd had a few feathers pulled from her back and was frightened, but other than that she seemed okay. She wouldn't let me catch her to look her over, nor would she come out when I scattered some scratch and called "chick, chick, chick." Couldn't say that I blamed her. I watched her for a little while to make sure she wasn't injured. When she finally got the courage to eat the scratch, she appeared to get around just fine.

After researching how to deter hawks, Dan decided to try a couple of things. There are numerous recommendations on the internet: scarecrows, mirrors, fake owls, mannequins, kites, shiny fluttery Mylar tape, fake hawks, shutting the chickens up till the hawks go away, big dogs, larger livestock, radios, roosters, netting, and hawk-call recordings. Human presence is supposed to help as well, although I once had a hawk try to nab a chicken within about ten feet of where I was standing.

We decided to use what we had on hand, which meant making a good old-fashioned straw-stuffed scarecrow. We used an old pair of jeans, an old sweatshirt and socks, a pillowcase for the head, and a baseball cap. Long sticks provided the form and straw the stuffing. We further embellished our scarecrow by taking five old CDs and attaching one, shiny-side up, to the top of the scarecrow's cap. I put the others in pairs, back to back, and dangled them with string from the scarecrow's arms. Chickens, goats, and cats alike were fascinated and underfoot to inspect the entire process.

When we stood back for a final inspection, we had a good giggle. If the chickens and goats weren't afraid of it, what would the hawks think? Dan said we'd probably find hawks perching on it the next day.

The critters were curious about the makings of our scarecrow.

If our critters weren't scared, why would the hawks be?

I've read mixed reviews on how well scarecrows deter hawks. Some folks report success, others don't. It seems to help if it's relocated frequently. In spite of that, we lost three chickens to hawks in the weeks that followed. We rerouted the chickens' grazing grounds to where there was more brush cover. The chickens, though, wouldn't leave their yard or the coop for days. After a week or so the chickens finally headed out to the pasture and the hawks seemed to disappear. We realized we must be in a migration route, and sure enough, the following spring we spotted more hawks.

The chickens' behavior changed after the hawk attacks. Anytime our rooster let out a cry all the hens would run for cover. We heeded his alerts as well, and would come running to check on things, glad he was more watchful than before. On another occasion, when I shook the dust out of a couple of throw rugs, the flapping sound sent the chickens running for the bushes. Obviously it reminded them of a hawk's beating wings.

In spite of our own efforts, the thing that seemed to help most was the flock of crows that moved into the area. I used to dislike crows because they were forever knocking over a former neighbor's trash can and scattering trash all over her yard. Plus they rob other birds' nests, which is not an endearing quality. However, crows dislike hawks, and when hawks are around, crows will chase and harass them. The crows will post lookouts at the tops of various trees and report back and forth to one another. Any time they all take off with a lot of cawing, we usually spot at least one hawk too.

Besides the migrating hawks, we have a few that live in our area year-round. They are territorial and with a literal movable feast on our place, it is difficult to dissuade them. Because of that, we realize we may occasionally lose chickens or possibly cats to these predators. It's more likely with free-range birds like ours, but even fenced chicken yards (unless they are covered with netting or fencing) cannot guarantee chickens won't be nabbed.

With a good rooster and the crows on patrol we haven't lost a chicken to hawks since. We heed the crowing and cawing, and scan the skies when we hear them, because I feel responsible for our chickens. In fact I sometimes wonder if letting them free-range is the best choice. If we lost our entire flock to predators I might consider other options, but for now, having the freedom of several large fenced areas enables them to be what they were created to be: happy, healthy, and going about their chicken business. They not only give us top quality, free-range eggs, but also help with insect control and eat quite a few weed seeds to boot. I just try to keep in mind that the balance of life as we know it, is death.

THE MYSTERY OF THE DISAPPEARING CHICKS

I am a habitual counter. I find myself counting things whether they need to be counted or not: steps on flights of stairs, the green beans I'm picking, or the number of whisk strokes it takes to beat eggs. Chickens and chicks do need to be counted, however, so when I kept counting twenty-four baby chicks instead of twenty-five I became concerned. They were still small enough to be confined to the coop, so I went from puzzled to increasingly worried with each counting. We'd never had a problem with our chicks before, so I was at a loss as to where the missing one had gone. No frantic peeping from a corner of the coop, no lifeless body, no sign, just disappeared.

After the second one disappeared into thin air a few days later, I became alarmed. The coop is fairly tight against intruders. It has a concrete floor, so no burrowing from underneath. It has three solid exterior walls and one interior wall of one-inch poultry netting. The windows are covered with half-inch hardware cloth. The building is shut up at night.

The only openings we could find were under the eaves of the roof. But what could fit through those narrow openings? We suspected a snake, because snakes will eat eggs and baby chicks. I started researching and found a homemade snake trap made from a minnow cage[1] so I bought two. We plugged the holes in the coop, set the traps near the back of the back of the building, and waited.

Although we didn't catch anything, all seemed well until a few days later. I had just finished the morning milking, and it was getting light enough to see inside the coop. I discovered a dead chick in the corner of the chick pen. Its neck was covered with dried blood. I began to count and realized three more were missing besides. We had to do something, but what?

By that time we'd lost six chicks. If our traps weren't working, the next best step seemed to be enclosing Mama Hen with her chicks at night, because that was when the chicks disappeared.

After racking my brain for a quick solution, I took a short length of welded wire fence, tied hardware cloth to it, and made a box to become a "bedroom" for mother and chicks. We also considered the possibility that it might be rats and set out rat traps. Then we plugged the holes in the coop, set a trap near the back of the coop, and waited.

Mama was none too pleased to be put into the safety cage each night with her chicks. In fact, she came up with her own solution:

Mama Hen's solution was to pile the whole bunch into one of the nest boxes. Considering that it wasn't very far off the ground, I was less confident than she was in its ability to protect her chicks.

I didn't trust this arrangement, however, and insisted they all go into that cage. My method for doing this was to herd them into it around dusk: still light enough outside to see, but dim in the chicken coop. I would put Mama in the cage, shine a flashlight on her, and all the chicks would run in after her.

One night we were about half an hour later than usual. I counted only eighteen chicks running into the cage. There were supposed to be nineteen, but they moved around so much that they were difficult to count. I finally gave up trying. As I secured the door, Dan found the dead chick near the same spot where the other had been. This one also had its neck bitten open. Needless to say Dan bought a live animal trap the next day.

It took three tries to get it, but we finally caught our chick killer: a rat.

Why it didn't go for the bait in the conventional snap rat trap was a mystery; we'd never before had a rat turn its nose up at peanut butter and cheese. This one, however, preferred raw chicken to grains, legumes, and cheese. We used a raw store-bought chicken neck as bait. It couldn't resist that. It wasn't heavy enough to trip the trap door the first night, so it made a clean getaway with the bait. The second night Dan added some weight to the trip plate, but it gnawed most of the chicken neck from the outside of the cage. Finally, Dan secured the neck with wire inside the cage so that the critter had to enter.

Our relief was brief, as we began to wonder if there were more of them. We continued putting up Mama and her brood, and setting the trap for a while, just to be sure, but that was the only rat we caught.

It wasn't long before the chicks were too big to be shut up in the safety cage for the night. They had begun to roost, but some of them (not surprisingly) refused to go into the coop at night and began to roost in the cedar tree in the chicken yard. During the next several weeks we lost two more young chickens.

I found the first new victim early one morning in the chicken yard, headless. Several of the youngsters had managed to spend the night in that cedar tree and something had plucked one out of the tree and killed it. I found the second one several days later, this time in the pasture along the back fence bordering the woods. It was mid-afternoon and I had just checked on the chickens a few hours before. This one was also headless.

Quite a few chicken predators consume only the head and crop. Most of them, however, hunt either at night or during the day. We appeared to have lost chickens during each of these time periods. Did that mean we had two different predators, a night hunter and a day hunter? I made sure all chickens were in the coop before closing up at

A rat had killed six of our chicks.

night. No more roosting in the cedar tree. I changed the chickens' free-range area, directing them to the front pasture with the bucks and away from the woods. Dan began to set out the live animal traps again. We also decided to expand the rodent control department and adopted two more kittens from the animal shelter.

Dan eventually caught two opossums, also known chicken killers, and disposed of them. The chicks were now too big to be attacked by rats, but the barn had become overrun with them, and we could not rest easy with the situation. I would see multiple rats in the chicken coop when I checked it at night, and Dan reported hearing them running in the walls.

Our chicken coop/goat shed was an old building, built about ninety years ago. It had gone through several transformations over the years. Apparently it was originally used as a chicken coop. At some point, someone poured a concrete floor and put up paneling to turn it into a workshop. By the time we bought the place it was a storage building. We used it to make a shelter for our critters and until now, hadn't had problems with rodents. Upon inspecting the building closely, we realized that rats had tunneled under the concrete floor and gnawed a series of entries and exits into the walls and all around the building. If we shined a flashlight down some of the holes, we could see pairs of beady eyes staring back at us, waiting for us to leave. Except for the chick killer we caught in a live animal trap, all the rest had deftly avoided all attempts to catch them. After searching YouTube, Dan built a rat-disposal system he saw on a video.[2]

A PVC pipe runs like a tunnel from a rat hole up to a PVC elbow which drops into a tub of water. A strip of old towel inside the pipe offers traction, and peanut butter at the mouth of the elbow is the bait. Rats fall into the tub and drown. The fellow who invented this said he successfully disposes of rats on an ongoing basis.

The next morning we approached this contraption with anticipation. The tub was full of water but empty of rats. Inside the chicken coop, however, I discovered a newly shed snake skin. We never did drown a rat with that YouTube rat trap, but thanks to that snake, which was able to slither through the tunnels under the concrete slab, the rat population decreased noticeably. Dan eventually found the snake—a rat snake—under the nest boxes. If it would only catch and eat mice and rats we wouldn't have minded it being around, but snakes also eat eggs and chicks, so Dan removed it.

We didn't want another rat population explosion; and with the snake gone, we feared if we let things be, we would soon be overrun again. It's difficult to control rodents in a chicken area because of open chicken feeders and loose chicken scratch on the ground. Plus, our rats knew how to avoid traps. Although we were extremely reluctant to do it, in the end we resorted to rat poison. We feared the cats would get into it, so we got a kind we could drop down the rat holes in the shed. Soon after, the poison did its job. We knew it was successful because of the terrible dead smell that filled the shed for quite awhile afterward.

Total loss from predation that summer was nine out of twenty-five chicks. I still continued to count chicks and chickens several times a day, always with a tinge of worry. Especially because we seemed to have conquered the rat problem just in time for autumn hawk migration.

That summer's surviving chickens never did quite get over it. They disliked going into that coop until we finally built a new one. We took precautions building the new coop, to make it as predator-proof as possible. Even so, protecting our critters is an ongoing process. We try to do our best, knowing that along the way we will lose a few. I reckon it just comes with the territory.

CHICKEN WRANGLING

I admit it: I am not an adept chicken wrangler. Fortunately, most of my chickens are pretty cooperative, although they get over the fence occasionally and have to be herded back. Calling, "Chick, chick, chick, chick, chickens!" and throwing a handful of scratch usually does it. The Silver Laced Wyandottes and Speckled Sussex will actually run up to the chicken gate when they see me and wait to be let in. But not my Buff Orpingtons. Whenever they see me, they take off in the opposite direction.

The oldest Buffs apparently never got over the trauma of my flapping the bath towel to herd them when they were chicks. The younger ones experienced the serial chick killer the following summer. They watched while the predatory rat picked off and killed seven of their number inside the coop during the night. Who wouldn't have psychological issues after something like that?

Ordinarily, not being able to herd my chickens wouldn't be a problem, but it became one recently. One of our homestead goals is pasture improvement. We try to pick one pasture area each year, have the soil tested, re-mineralize it, and plant it. Chickens become counterproductive when they invade the newly planted pasture and eat all the seed. Once the new forage begins to grow, chickens are notorious for scratching up everything. I didn't see how I could get a new pasture established with the chickens around. There is plenty around the homestead for them to eat without undoing all of our hard work, so I began to work on ways to keep them out.

The first thing I tried was to reroute them into another forage area. I closed all the gates and covered any chicken-sized openings. That stopped all but four persistent Buffs. I observed that most of the fence-hopping occurred at the cross members of the H-braces, so I used baling twine to create a barricade. This deterred them somewhat. They would make a terrible fuss when they jumped up and had to negotiate the twine, but it didn't stop them.

As I racked my brain, I thought of our former hawk deterrent efforts. We used the shiny side of old CDs to decorate our scarecrow, because shiny things are supposed to help keep hawks away. Maybe it would work for chickens too. I didn't have any spare CDs, so I attached a row of aluminum foil strips to the top of the fence. Aluminum foil is well known for its anti-alien properties[1] and I figured that if this worked, I could make my fame and fortune by proving that chickens are aliens in disguise from another planet.

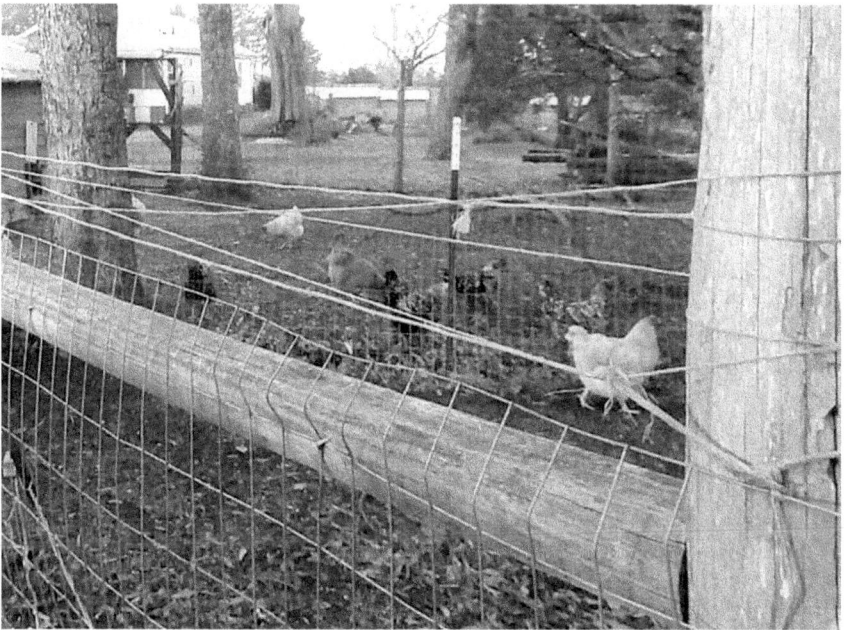

Cross members (cross braces) are places where the chickens like to hop the fence. Baling twine made an easier barricade, although it wasn't impermeable.

Alas, this did not work. Those four stubborn Buffs still managed to jump the fence in the morning and work that pasture. What made matters worse was that they couldn't figure out how to get back into the chicken yard. I would try to herd them, but they insisted on running away in the opposite direction.

This went on for several frustrating days until at last I decided they had to be caught and wing clipped. You'd think this would have been my first solution, except that these four didn't go into the coop at night to roost. Rather, they roosted at the top of the cedar tree in the chicken yard. Now, however, they couldn't get back and were roosting in the bushes bordering the pasture I was trying to establish. I figured I could catch them at night, clip their wings, and all would be well.

I thought I would have a good chance to catch them if I waited until after dark and nabbed them unaware. Except they weren't unaware; they saw me coming in the moonlight. After freaking out the first two hens, I went after the other two with the flashlight. It blinded them so that I was able to pluck them out of the bush easily. Then I went after the getaways, which were also blinded by the bright light and simply hunkered down. That made them easy to catch, too.

All four had both wings clipped and were placed in the coop that night. At last I was able to complete my planting and give the pasture time to establish itself. As an extra precaution, so that no chicken would be tempted to try to get into that pasture again, I did the planting at night by the light of the moon. I figured if they didn't see me throw down the new seed, they wouldn't think about trying to break back in. The chickens were none the wiser and the pasture grew. Success.

MY PERSONAL CHICKEN

When I chose my first chicken breeds, one of them was the Ameraucana. I chose this breed because of the eggs they lay: beautiful tints of green, cream, or blue. I thought that since this was Dan's first experience with chickens, he would enjoy the rainbow of egg colors that our four breeds would lay. Of my six Ameraucana chicks, only two turned out to be pullets. One laid a pinkish cream-colored egg, but the other laid light green eggs.

Like my other chickens, the green-egg layer never had a name, although Dan took to calling her "The Ugly Chicken." Nonetheless, she appointed herself as my own personal chicken. Not that I wanted a personal chicken, nor that I think of chickens as pets. I don't. It doesn't mean I don't like chickens; I love chickens. Given our lifestyle, it just doesn't make sense to become emotionally attached to our animals. This chicken, however, seemed to attach herself to me. In about the middle of the afternoon, she would come looking for me. I'd look outside the kitchen window and there she would be, at the bottom of the porch steps, looking up at the house and waiting for me to come out.

If I was already outside, she'd come running right up to me. She'd stand at my feet and cock her head to look at me the way chickens do. Then she'd cluck as if to say, "I found you." After that she'd follow me around. In the event that she didn't find me, she'd busy herself by scratching up anything I'd recently mulched.

According to homestead rules, she wasn't supposed to be on this side of the fence. She was a fence hopper though, and defied every wing clip I gave her: one-sided, both-sided, even, uneven, lopsided, you name it. She still made it over the fence.

One autumn day she saw me raking leaves, hopped the fence, and came running. She happily scratched down all the leaves I was trying to rake up. I don't deny that this was a tad annoying. As friendly as she was, however, she wouldn't let me catch her.

The Ameraucana is one of three recognized breeds to lay green and blue eggs. The other two are the Araucana and the Cream Legbar. Crossbreeds of these can also lay blue or green eggs and are lumped together as "Easter Eggers." Ameraucanas are often confused with Araucanas and Easter Eggers, so to avoid that confusion, the Ameraucana Breeders Club has strict guidelines in its American Standard of Perfection.[1] The biggest difference between the Araucanas and Ameraucana is that Ameraucanas have tail feathers, while the Araucanas are "rumpless" or tailless. True Ameraucanas sport pea combs, small wattles, facial tufts, and slate blue legs. Easter Eggers may have some or none these characteristics. Ameraucana colors and markings can vary considerably.

My two Ameraucanas laid some of my largest eggs. They also laid through winter, which meant we were never completely eggless. They laid well through their second year, but at about three years old, my green-egg layer began to produce less and less. By that time she had rather distinguished herself around the homestead, which in turn presented a dilemma.

This dilemma presents a struggle for almost every new homesteader, especially those who choose to raise their own meat. Most of us come from typical modern backgrounds, where one's animals are seen as pets. The novelty and delights of those first chickens or first goats leaves an emotional imprint. We can choose in our heads to raise these animals for meat, but the heart seldom follows the head. It's usually the other way around.

I waited many, many months in hopes of more green eggs. When it became apparent that there would be no more, I'm sure many would

One of my two Ameraucana hens, the green-egg layer, became my personal chicken. Every morning she would wait for me at the bottom of the back steps.

have allowed her to live out her retirement years on the homestead. This is often an emotionally satisfying decision, but there are other considerations. Obviously a nonproductive animal requires feed, water, and living space, just as productive animals do, but not necessarily equally so. Older animals have risen to the top of their species' social order, so that they get first dibs on food and treats. The younger, productive animals, who need the nutrition most, often get chased away or otherwise pushed aside, so that they get less or none.

As animals age they often have more health problems, just as people do. That means they require more care, along with the decision to either pay a vet or treat ourselves. I've found that most health problems I choose to treat myself require extensive research to diagnose and understand how to treat. Medicines cost money and treatments take time and energy. Our emotional attachments may deem that it's "worth it," but it also means less time, energy, and money for the rest of the homestead. If all I was doing was raising chickens that decision may have been acceptable, but if the goal is self-sufficiency, like ours is, then there is never enough time, energy, and money to go around even when everyone is healthy and things are going well. What happens when the scales are tipped in favor of any one thing or animal?

Obviously other things suffer, and I can tell you from experience that eventually this kind of situation becomes frustrating.

The choice about what to do is an individual one and it's often an emotional struggle. The truth is that we can never really save anything. Everything eventually must die, because without death, there would be no life. What has helped us is our commitment to our prime directive, regardless of how we feel. If we want to become as self-reliant and self-sustaining as possible, then everything we do must conform to those goals. We decided we would keep chickens for eggs, meat, manure, and reproduction. I would raise chicks every year to replace our oldest chickens. There is no burden of decision-making because the goal is the decision. And as tough as it seems, it's actually a relief once all is said and done.

You've probably guessed by now that we decided that the next time we processed chickens, this hen would be one of them. The amazing thing was that she somehow managed to pop out of the killing cone and escape. She had a reprieve, but the next time, Dan had a better grip and all of my oldest hens were used for a batch of canned chicken. Even in death they had purpose.

MOVING DAY FOR CHICKENS

When we first started looking for property, we made a wish list of everything we hoped for on our dream farm. One of the things on that list was a barn. The place we eventually bought did not have a barn, but it did have two old outbuildings, one of which we thought we could modify to house chickens and goats.

I think there is much to be said for making do with what one has available. Even though I had done my homework regarding chickens and their needs, it was the day-to-day tending to them that helped to shape my idea of a better chicken coop. We tried to think things through when we planned for chickens and modified that old shed, but once we were in the thick of it, there were problems I hadn't anticipated. For example, I didn't have anywhere in the first coop for a broody hen and baby chicks. Something else we didn't anticipate was the arrangement of the doors. Our set-up was such that I could not get a wheelbarrow into the coop, which made it very difficult to clean out. The problem with the rats was unanticipated but very much factored in to how we wanted to design our next chicken coop.

We looked online for plans and discussed ideas of our own. Eventually we found something we liked in Carol Ekarius's *How To*

Build Animal Housing.[1] It was a plan credited to Penn State Cooperative Extension[2] and was about the size we wanted. I liked it because it included a storage room. With our goal of trying to raise all our own animal feed comes the need to store it all as well. However, we never seemed to have enough storage space and our small outbuildings accommodated very little else besides the animals and a couple grain sacks' worth of feed.

We did make a few modifications to that plan. Besides enlarging it a bit, I wanted to change the arrangement of the doors. I wanted a straight shot into the coop through doors wide enough for a wheelbarrow. We also moved the location of the exterior door based on the arrangement of our own barnyard.

Other things I considered included placement of the roost, nest boxes, feeder, and waterer. Also I wanted to allow enough room for a baby chick brooding area. Another thing we wanted was a generous, porch-like overhang to protect the entrance from rain and snow. While I worked on these plans, Dan focused on location, the foundation, and roof designs. See "Resources" for web links to more information on our design and building details.

The chicken coop was also important because it was the first step toward building a goat barn. By building the coop first, we could use the old coop for storage while we built the new barn.

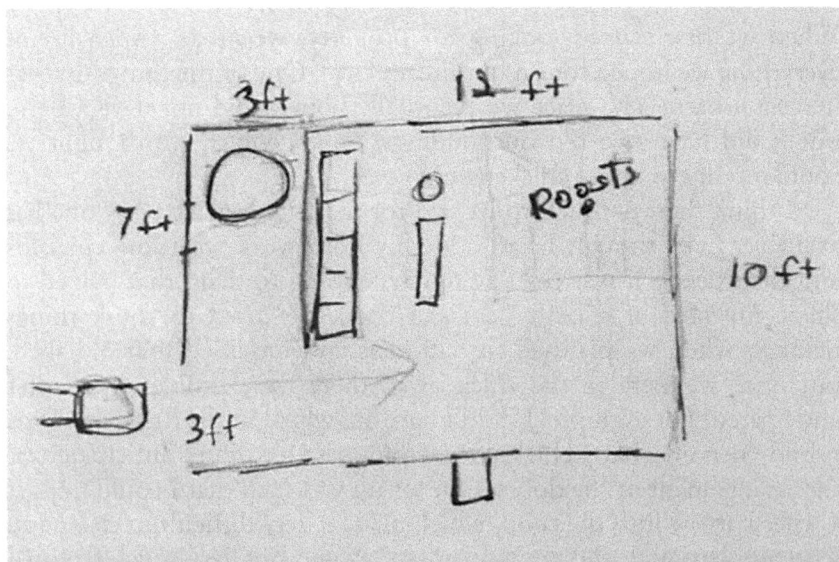

My rough sketch of our new chicken coop. I probably should have made a more exact sketch on graph paper, because the measurements of the finished coop are a bit different, but the basic idea remained the same.

The new coop sports a barn door entry and recycled storm windows.

Ground breaking for the new coop was in February, and it took about three months of part-time building to complete. I couldn't have been happier with it when it was done. It's roomier and brighter than the old chicken house, with easier access for both chickens and humans, plus it had that storage space.

The next step was to move in the chickens. Because chickens (like other animals) do not like change, I put some thought into how I was going to do this. I finally decided the best thing to do would be to move them at night. I'd give each chicken a wing clipping, put them in the new coop, and leave the flock inside their new quarters for at least two nights with a full day in between. When we allowed them out again the door to the old coop would be closed. I hoped that by doing it this way, they would adapt more quickly to their new home.

What complicated matters a bit was that I had a broody hen in the old coop setting on a small clutch of eggs. I worried that if I moved her she would abandon her nest. Still, we had to get the chickens moved so this had to be a risk we were willing to take.

To let the chickens become somewhat accustomed to their new coop before moving day, we opened its chicken door and let them have a look around inside. I also moved their feeder into the new coop, so that they would have to go there if they wanted chicken feed. We left things like this for several days before the move.

What did the chickens think? They were curious but cautious at first.

I had to lure them inside with chicken scratch.

The chickens seemed to like it, and inspected everything from the new nest boxes to the fresh lawn clippings on the floor. The interior was painted with old-fashioned, home-made whitewash. Made from hydrated lime (calcium hydroxide), it is said to help control insects.[3]

The tree roost was an idea which we thought was clever, but for some reason our chickens had trouble negotiating it, and some refused to use it. We later replaced it with a conventional type of chicken roost.

On Chicken Moving Day, I waited until after dark to commence the move. Chicken by chicken, I removed each from its perch in the old coop, clipped its wings, and moved it into the new coop where my Coleman rechargeable camping lantern lighted their way to their new roost. The wing clipping was necessary because, besides not wanting chickens scattered everywhere and getting into everything, I wanted all of them to lay their eggs in their brand new nest boxes. I was tired of their fence hopping and my daily hunt for eggs. I wanted to collect the eggs all in one place, where they were supposed to be!

The last to be moved was my broody hen and her clutch of five eggs. I carefully removed the eggs from underneath her and placed them in a basket. Then I carried her and her eggs to the new coop, where a nice nest of fresh straw awaited them. The eggs went into the nest first, and then mama was set down next to them. She was mighty upset about the whole thing! After a huge fuss, she finally settled back down on her eggs.

The following morning I went to check on the chickens as soon as day broke. No chicken seemed unduly upset, and they clucked and scratched around like it was Old Home Week. Mama sat patiently on her nest as though nothing had happened.

There was actually a waiting line for the nest boxes. I couldn't believe it!

After two nights in the new coop, I let the flock out and into their newly expanded chicken yard. Every chicken acted as though nothing had happened until dusk, when a little confusion presented itself. When I went to shut up the coop for the night, I found five hens and the rooster waiting dutifully by the old coop door. I had to catch two, but then the others ran into the new coop for the night.

And mama hen? Within the week she hatched out one egg. The other eggs didn't make it. But she was just as proud of that one chick as we were of our new chicken coop.

THE WEATHER CHICKEN

Of memorable chickens, I can't help but include The Weather Chicken. And what, you ask, is a Weather Chicken? Well, this particular chicken, a Silver Laced Wyandotte, managed to escape a bout of wing clipping and continued to roost in that cedar tree in the chicken yard. With clipped wings the others had to roost closer to the ground, i.e. inside the coop. But not The Weather Chicken. She steadfastly roosted in that tree all by herself. The exception was when the weather was going to be snowy, sleety, rainy, fiercely windy, or otherwise nasty that night. She always seemed to know and would opt to roost in the coop on a bad weather night. Amazingly, she was always right!

That cedar tree in the chicken yard was a favorite roosting place for quite a few of my hens. Even in winter, when temps dipped below freezing at night, I had that traumatized core group of chickens which stubbornly refused to roost anywhere else. This was worrisome when there were winter storm warnings. I had to add "get chickens out of tree" to my storm chore list:

Winter Storm Chores:
  fill kindling boxes
  cover wood pile
  put away tools
  fill hay feeders
  fill chicken feeder
  fresh bedding
  fresh warm water in all buckets
      and waterers
  buckets in insulation tubs if
      necessary
  get chickens out of tree
  check and fill kerosene lanterns
  get a nice roast out of the
      freezer (enough to feed us
  for a couple of days should the electricity go out)

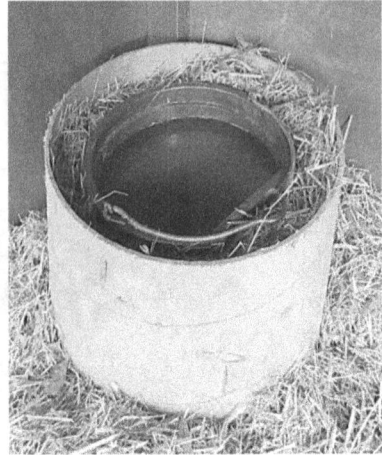

Straw insulates a water bucket.

This particular chicken chore wasn't too bad at first, because they all roosted in the lower branches of that tree. It was easy to pluck them off and pop them through the chicken door of the coop. They soon caught on to this and began to roost a little higher. I found I could take a long garden tool, such as a rake, and jostle their roosting branch until they flew down by themselves. If I did this after dark and shined a flashlight on the chicken coop door, they would invariably run in, fussing all the way.

I don't know if it was breed smarts or if this particular chicken was one smart cookie, but The Weather Chicken took to roosting in the very topmost branches of that cedar tree, making her impossible to remove. So I let her be and enjoyed a personal heads up on the overnight weather.

Eventually we had to catch her and clip her wings too. We had just completed our new chicken coop and I planned to move the chickens from the old coop to the new after dark. I wanted to give them a fresh wing clip because the flight feathers grow back.

When it was time to move the chickens into their new coop we decided she'd have to be caught too. Dan used a stick to jostle The Weather Chicken's perch in the cedar tree. She flew down and ran into the old coop. It was easy to catch her then. She had her wings clipped like all the others and was put into her new home.

My three Wyandottes were nearly identical, so that I no longer knew which of them once distinguished herself as The Weather Chicken. Still, she'll never be forgotten.

RULES WITH AN IRON CLAW

Dan called him "The Sultan" because he kept his little harem of Buff Orpingtons in check. He was the third reigning rooster we had on the homestead. It's interesting how different each of the three has been. They've all done what roosters do, but not exactly the same way.

Roosters are not necessary for hens to lay eggs, but we decided to keep one because we want to raise our own chicks. Some people decide not to keep them, and for some pretty good reasons: they crow at all hours and drive the hens crazy, because they are relentless when it comes to heeding the mating call. It can become so bad that the hens soon sport featherless backs. This is why it's a good idea to have at least ten to twelve hens if one keeps a rooster. In addition, it's not uncommon for roosters to be aggressive and downright mean.

In their favor, besides producing chicks, roosters are good watch guardians of their flocks. They remain alert and watchful while the hens scratch and forage. When they find something they think the hens would like to eat, they'll cluck to call them over and then step back to let the hens enjoy. Because of their protective instincts, they'll sound an alarm whenever there is potential danger. If our rooster sees or hears a hawk overhead or a stray dog in the yard, he'll alert the hens who will run for cover.

Our first rooster was Lord Barred Holland (or Lord B), so called because of his lordly manner, although Dan simply called him B. He was almost perfect: always watchful, always on the alert, deferential toward his hens, and non-aggressive toward humans. At least that was initially the case. As he got older he started to show signs of aggression toward us, which was absolutely not allowed. His biggest fault, however, was his intolerance of new additions to the flock, both pullets as well as cockerels.

Our second rooster was Cowboy, a Buff Orpington. Cowboy did not seem to have all the natural roostering qualities. Mrs. Mean had to teach him that any goodies he found were for the hens, not himself. I don't know how many times she had to rush in and gobble down his food-finds until he learned to simply announce them and step back to let the ladies enjoy. But he did catch on. He was a good watchman, if not an alarmist. I sometimes suspected he liked keeping things stirred up and I have to admit the place was a lot quieter once Cowboy was no longer with us.

The Sultan showed much to recommend him as a worthy flock rooster, along with a few qualities that didn't. He was extremely alert and watchful, but he ruled with an iron claw. When he first realized he was the only roo, he chased down each of the hens in turn and simply stood on top of her. Eventually it got to the point where any time he approached a hen, she would hunker down in obeisance before he grabbed her. This was mostly true of the Buffs, because the Wyandotte and Speckled Sussex hens had minds of their own. They went where they pleased and as fast as they could if The Sultan was on their tail. Those Buffs, however, could almost always be found in a cluster with The Sultan standing tall and proud in their midst. To his credit, he

The Sultan has been our only rooster so far who left the chicks alone. That is, until they grew up.

respected the humans and kept out of the way. He was very alert and the only rooster we've had that seemed to be able to round up the hens if he heard a hawk or saw one fly overhead. The hens heeded and obeyed, and I didn't lose one chicken to hawks during his reign.

The Sultan was the first one out of the chicken coop in the morning. As soon as I opened the chicken door, there he was, strutting his stuff. The hens all held back. They knew that the first few out the door would be required to pay homage to the lord and master of the flock. He particularly seemed to like the Buffs and would chase them first. I don't know if this was because Buff Orpingtons are more docile as a breed, or because he preferred blonds.

One excellent quality about this rooster was that he did not bother the chicks. In fact, he didn't seem to take notice of them. Considering our previous troubles with roosters, chicks, and flock integration, this was a reason I wanted to keep The Sultan around, even if he was something of a bully.

The only two chicks that hatched that summer were cockerels. Some folks seem to do well with more than one rooster and, considering how accepting The Sultan had been of the chicks, I wondered if he would be as tolerant once they grew up.

When the two young roosters grew old enough to look like roosters, The Sultan started chasing them away from the hens, food, and coop. Initially, the youngsters kept their distance from one another, but eventually they paired up. Every morning, The Sultan would chase them behind the coop where they would hide until the chickens were let out to free-range. Then, they would go back into the coop, wait for a hen to come lay her egg, and jump her. This would set the hen into a panicked flurry, which always brought The Sultan running in a fury.

As time went on, one of the cockerels grew bolder and eventually started standing up to The Sultan. When The Sultan started backing off, we knew that a challenge for the top spot would be going from a simmer to a full boil in a matter of time. The crowing, challenging, and chasing kept getting worse, until we knew we had to do something. We decided to cull all the males. That included The Sultan, who protected, but also bullied, the hens. The hens are the producers on the homestead and we do not want them to be harassed. We would be roosterless for a while, but I felt the hens needed a break. Plus, they were all getting older, which meant we needed to start thinking about adding pullets to keep up egg production. I figured we'd find a replacement roo when we got our new batch of baby chicks.

In spite of being somewhat overbearing with his flock, The Sultan was a pretty good rooster. Still, no rooster's reign lasts forever.

VOLUNTEER DAY ROOSTER

We didn't remain roosterless for long. That wasn't because we deliberately got another rooster; rather, the junior rooster next door was quick to notice that our flock of lovely ladies had no male companion. He quickly made himself available for the job.

The first time I realized he was there was one morning when I went out the back door and heard a rooster clucking his prettiest "I found a treat to eat" song. I looked and there he was, standing outside the chicken yard, trying to coax half a dozen of my hens to come see what he'd found. They were extremely interested but were on the other side of the fence.

The next morning, I found Mr. Rooster in the front pasture with the hens, just as happy as he could be. Every morning after I let the chickens out, he would show up and spend his day with the hens, who certainly didn't discourage him. I have to admit he was much more gentlemanly than The Sultan had been, but that didn't mean I wanted to invite him to stay. Our next door neighbors seemed to take no notice, but I was sure their senior rooster didn't mind getting rid of the competition.

Initially, he would meet up with the hens after they'd been let out for the morning. Gradually, he began to follow them back to our chicken yard, but he couldn't quite figure out how to get into it. This was because of the arrangement of the pastures, gates, and yard. When going out, the chickens had no trouble finding the gate leading to the front pasture. When heading back, they would often bypass that gate and head for the corner nearest the chicken yard door instead. They had to remember to go through the gate first to get to their yard entrance. The bright ones did, but a few of the others ended up in the corner and didn't "remember" until they saw another chicken make it successfully back into the yard.

Our volunteer day roo was usually around at chore time, so Dan or I would toss chicken scratch over the fence to him. Even so, no one, including the hens, went out of their way to show him the secret entrance to the chicken yard. During the day, he happily escorted the hens around the pasture, then headed back home for the night. This was fine with me because I really didn't want him taking up permanent residence. What would happen when we got new chicks, and young roosters began to grow up? I didn't know. We'd have to cross that bridge when we came to it.

Roosters are an important part of chicken society. Without one, a hen will sometimes stop laying to guard and even try to crow.

Sadly, our volunteer met his demise before we got to that point. One morning I found him in the front pasture with his head chewed off. Dan thought a 'possum had gotten him, and several days later we caught it in the live animal trap. Shortly after that, all the chickens next door disappeared, and no other volunteer rooster came to call. Instead, one of our hens took up trying to crow, something which is not uncommon in flocks with no rooster.

Our next rooster will come from our next batch of mail order chicks. Until that time, we will remain roosterless.

OF CHICKENS, GOALS, AND WHAT I'VE LEARNED

Our goal from the beginning has been self-sufficiency. By that I don't mean isolating ourselves from the world, but rather, not being dependent on the industrialized consumer system to meet all of our needs. Self-sufficiency to Dan and me means meeting more of our own needs by working toward becoming as self-reliant and self-sustaining as we can. It is an ongoing process. Regarding chickens, I've learned quite a bit through the knowledge and experience of others, as well as my own; but, especially, I've learned from the chickens themselves.

Working toward sustainable chicken keeping is a partnership between me and my chickens. It has meant learning how to provide for them from the work of our hands: feed, water, pasture, shelter, protection from predators and disease, plus an understanding of how many chickens our land can support. They, in turn, provide eggs, meat, manure, chicks, and some degree of insect control. They peck through the manure of the goats, hunting for undigested grain, parasite eggs, or fly larvae. They do a pretty good (though not always welcome) job of spreading and turning organic matter on the soil surface. Finally, their entertainment value is priceless.

How far along are we in accomplishing our goal? I'd say the biggest challenge (one we're still working on) is feed. We are able to offer quite a bit for our chickens from what we produce on the homestead, but not everything—yet. Besides what they forage for themselves from the pasture and woods, they get scraps from garden and kitchen, and whey from cheese making. We have been able to supplement purchased grains with what we've grown here: wheat, corn, cowpeas, amaranth, grain sorghum, and sunflower seeds. Even so, we still have to purchase most of it. Because our chickens free-range, feed doesn't need to be "complete" in the sense of providing every nutrient they need, but it does need to be practical as well as simple.

There are numerous chicken feed recipes out there, most of them with a lengthy list of expensive ingredients which may or may not be readily available. A simple recipe which uses things we can realistically grow or buy locally is

> 1 part wheat berries
> 1 part corn (if homegrown) or oats (if purchased)
> 1 part small-sized legumes, such as lentils or cowpeas

If I'm able to use homegrown grain, I use corn because it's easiest to harvest and process. If I have to buy it, I opt for oats, because most of the corn raised and sold in the US is genetically modified. I choose not to use that. Also, small legumes can be difficult to find in bulk at a reasonable price, which means I am not able to use this recipe as often as I'd like if I don't have homegrown cowpeas.

This feed mix should give an approximate 16% protein content, which is recommended for laying hens. I calculated this using the Pearson Square.[1] I confirmed it by having the mix tested through our cooperative extension service. This can provide the backbone of our chickens' diet, which is supplemented as I mentioned above. Feed is left out free choice, and I can judge how well the free-ranging is going by how quickly the feed is eaten.

The challenge is growing enough grain and cowpeas for a year's supply. For example, if I buy chicken feed and scratch about once a month at 50 lb. each, then between the two I'm feeding my chickens 25 pounds per week. That means I must grow, process, and store 1300 pounds for a year's worth. That would be approximately 434 pounds of each ingredient for the chickens alone. Here's the do-it-yourself nitty-gritty on that.

GROW: So far our grain-growing has been mostly experimental with a lot of disappointments. We've done best with corn, wheat, and amaranth, but oats and barley have not done as well. Other things I've tried include various sorghums, flax, and sunflower seeds (usually gobbled up by song birds). Most of these have been experimental patches, helping me determine whether it can be grown here and how much yield I can expect from a particular size of plot. Being able to grow a year's worth is a goal I am still working toward.

PROCESS: Modern livestock feeds are pelletized for easy packaging. It is possible to purchase a pellet-making machine, but they cost thousands of dollars. Also grains are often cracked for feed mixes and chicken scratch, for which a hammer mill would be needed. But is cracking grain necessary? In some ways it makes sense to crack grains, making them smaller and "easier" to digest. That may be true for grazing and browsing ruminants, which can't really digest grain anyway (whole grains often pass through them whole, and the chickens will eat those grains from the manure). It also makes sense for chicks. For adult chickens, however, is it necessary? Opinions vary, as can the size of the corn kernels depending on the variety of corn. This is why I chose Truckers Favorite, a small-kerneled field corn. My cowpea is another small-size variety called Ozark Razorback. By avoiding large varieties, our feed can be eaten readily by my adult chickens without the extra equipment and step of cracking the grain. In addition, whole grains store better than cracked without loss of nutrients, another plus.

A cast iron hand-crank corn sheller makes easy work of processing corn.

Something else my chickens have taught me is that not all things need to be processed, i.e. wheat doesn't have to be threshed, nor cowpeas shelled. Chickens will pick out the kernels and peas just fine, although the dried pea shells can be fed to the goats for both bulk and roughage. More on that when we get to the goat tales. Of amaranth, I can toss whole heads into the chicken yard for them to peck away at. All these things save work and time, both valuable items on a working homestead. The one thing I do have to shell is corn. I do this with a hand-crank corn sheller.

STORAGE: Okay, let us continue with the previous example of 25

Storage room in the new chicken coop. Storage is one of our biggest self-sufficiency challenges.

pounds of feed and chicken scratch per week. A 32-gallon trash can will hold about 150 pounds worth of feed. So, for that year's supply of 1,300 pounds I would need 8.66, let's make it 9, trash cans. Are you mentally trying to picture where you would keep 9 trash cans of chicken feed? Me too! Another option would be 55-gallon, food-grade drums. These come in metal or plastic. I estimate they could hold more than 250 pounds worth of grain, in which case I'd only need 5 or 6 of them. Getting the last of the grain out of the bottom of them would be another consideration, but we won't worry about that at present.

If I choose not to process my wheat and cowpeas, the bulk of the husks and shells adds to the amount needing to be stored. This is true of corn as well, which I will eventually shell. As you can see, saving in one area costs something in another.

I should add that storage needs to include protection from varmints (any critter one doesn't want around is a varmint: mice, rats, squirrels) and insects. Cats and rodent-proof containers help with the varmints, but insects are the hardest to keep out. Pantry moths thrive in my part of the country. They don't eat the entire grain, but the larvae burrow into it and eat the germ—the most nutritious part. The starches are still of value, but my critters need the whole grain, not just the starch. I find that chunks of cedar branches in the grain help deter moths and other insects.

Recently I learned of a grain-free method of feeding chickens from a video by Australian permaculturist Geoff Lawton.[2] Chickens are given free access to huge compost piles containing barn cleanings and food scraps. The example in the video used restaurant waste, so the amount of food scraps was tremendous and the results and quality of the compost were impressive. The chickens constantly dug through the pile, feasting on the scraps, insects, earthworms, and whatever else they could find.

The appeal of this to me was how it partnered with the chickens, utilizing their natural behaviors to feed them while producing a valuable product in the process. It is a work-smarter-not-harder method for needful things on the homestead. I set about experimenting with this method, even though I cannot provide the amount of food waste shown in the video. My garden, kitchen, and food processing waste go to goats and pigs as well as chickens. Adding the chore of collecting and hauling restaurant scraps would mean less time to do other needful things. After spending years trying to keep chickens out of the compost piles, I decided it would certainly be easier to let the chickens eat from them! Even if this method only supplemented their diet, it would be beneficial.

Our three-bin chicken compost pile has a cement board back and cinder block sides. Boards in front prevent the chickens from scattering the compost too far.

A dozen heritage breed chickens remains our goal. These Black Australorp chicks will hopefully fulfill that plan nicely.

The last consideration is: how many chickens can our land support? There is no mathematical equation or formula to determine this, because many factors go into the answer. The number of chickens we can keep may vary depending on time of year and weather. Summer forage which receives abundant rain will certainly support more critters than hot, dry pastures that receive little or no rain. A mild winter will offer more than a severely cold one. Other factors are housing and feed stores.

Although my original goal was to keep a dozen chickens, I have, at times, kept closer to two dozen. This includes older hens which lay few eggs, hens in their prime laying years, and pullets which aren't laying yet. This gives me a good number of eggs—about five or six dozen each week. We process chickens once or twice a year. That means we don't eat a lot of chicken but we do eat a lot of eggs.

Their toll on the land is seen in pecked-down plants and scratched soil. In return they deposit manure. The bare areas become the target for spot seeding as needed. They also help in fly control by scratching up the bedding in the goat shed, looking for larvae to consume. We've thought from time to time to add more chickens and sell eggs on a regular basis, but I honestly don't think the land would do as well.

The bottom line is that if our chickens are well cared for, they are healthy and produce high-quality eggs, meat, manure, and more chicks.

EGGS: As egg-laying fluctuates throughout the year, so does our egg-eating. I have learned that we have to adapt our diet to what our chickens provide. I mentioned previously about freezing eggs, but there is a way to store eggs without electricity, using water glass (the common name for sodium silicate). Once commonly available at hardware and drug stores, it is hard to find nowadays. The water glass is diluted with water: 10 or 11 parts water to 1 part sodium silicate. Fresh, clean, unwashed eggs will keep for four to six months in this solution.[3]

Water glassing is an old-fashioned preservation method that still works today.

Eggs can also be stored by pickling. This is an excellent, tasty way to keep hard-boiled eggs and use leftover pickle juice as well. Just pop the peeled hardboiled eggs into a jar of pickle juice. We like juice from dill pickles the best and sweet pickles the least, but it's fun to experiment. Pickled eggs are tasty by themselves or chopped for egg salad or any kind of salad.

MEAT: Although you might not recall who said it, many of you are probably familiar with this campaign promise: "A chicken in every pot and a car in every garage."[4] The line comes from a 1928 speech by Herbert Hoover, promising prosperity to all if elected. We might not think much of this today, because thanks to industrialized chicken production, chicken is an inexpensive and common meat nowadays. Now that I raise my own chickens, I better understand the appeal of that promise. Perspective changes if one wishes to practice sustainable chicken keeping on the homestead.

One beef cow or a pig can supply a year's worth of meat for a family. One chicken, depending on how it's prepared, yields roughly one meal for a family of four, and two or three meals for just Dan and me, depending on how I stretch it. To eat chicken three times per week, that family would have to raise, process, and have freezer space for 156 chickens for a year's supply. For Dan and me, it would mean only about 52 chickens or so. You can begin to see the logistical problems with this, including having enough eggs to successfully hatch that many, then feeding, housing, raising, and processing all those birds. And that's not counting chickens for producing eggs for home-use plus all those chicks! Then too, more chickens means raising and storing more feed, along with larger housing facilities and the job of keeping those facilities clean.

Many folks purchase broiler chicks to raise for eating, which is an excellent way of knowing that one's food is being properly nourished and humanely raised. Unfortunately, broiler breed adults are not inclined toward egg laying, hatching, and mothering. Production layers are a little skinny in the meat department. Actually, dual purpose chickens are too. To eat only homestead-produced chicken likely means eating less chicken.

MANURE: According to J. I. Rodale's *Encyclopedia of Organic Gardening,*[5] an average-sized hen produces about 140 pounds of manure per year. Multiply that by the number of chickens one wishes to keep and, well, that's a lot of manure.

Visitors to the homestead are often surprised that my coop doesn't "smell." When they ask how often I clean it out I say once or twice a year, a second surprise. Chickens stink; everyone who has driven by a chicken farm knows that, right? But this isn't inherent with chickens; rather, it is the consequence of management techniques. Where hundreds or thousands of chickens are tightly housed under one roof, manure and urine accumulation become a problem. Huge fans are used to ventilate these buildings not only for the smell but for things like the ammonia from urine, which is harmful to breathe.

Maintaining smaller numbers of chickens in roomier housing with free access to pasture avoids these problems. I manage manure and urine with the deep litter method. This technique uses the same ratio of nitrogen to carbon as compost—about 1 to 30. By using six inches or more of carbonaceous material as manure and adding as needed, there is no smell in the chicken coop. If it does begin to smell, I stir it up (to add oxygen), then add more carbon: dried leaves, chopped straw, pine needles, grass clippings, etc. Every day I toss in a little chicken scratch so that the chickens will keep it stirred up for me. A couple times a year I clean it out and use it for sheet composting. Simple as well as effective.

For composting, chicken manure is considered a "hot" manure because it is one of the highest in plant nutrients, particularly nitrogen. What that means to me is that if my compost pile is a bit slow in the making, it takes less chicken manure to get it kick-started than goat manure. It also means it isn't a good idea to apply fresh chicken manure to the garden, because it will likely burn up the plants. With

Some folks say using a hen to hatch chicks is less predictable than using an incubator, but mother hens make it easier by doing all the work themselves.

the deep litter method, by the time I clean out the coop, the manure is decomposed enough to use for the sheet composting I mentioned.

REPLACEMENT CHICKENS: Throughout my *Chicken Tales*, I've described my experiences with hatching and raising chicks. My personal conclusion is that the right rooster is key to success. I should add that I am fortunate that neither predators nor disease have ever wiped out my entire flock, and we have managed to maintain an adequate number of chickens to meet our needs.

In the end, the key to sustainable chicken-keeping is balance. Like everything done nature's way, results are neither consistent nor guaranteed. That can be worrisome if one is looking for predictable results, but Dan and I came to realize that this is inherent in the lifestyle we've chosen. Rather than trying to determine a desired outcome beforehand and working to ensure production results, we are the ones who must adapt to what our critters and land produce. This requires flexibility and faith. It means setting aside the modern economic concept whereby increase and growth are the only good. It means understanding that bigger is not necessarily better and that accumulating more and more means more waste. It means a willingness on our part to tighten our belts if necessary without losing our sense of gratitude for our land, our critters, and our relationship with them. Like everything else on the homestead, this is a learning process.

Goat Tales

BUT FIRST, FENCING

Anyone who has goats, or knows someone who has goats, likely has a story to tell. Just as likely, the stories will involve fences. I am no exception and you will run across mine as you read through my Goat Tales chapters. It's not unheard of to have a goat jumping over fences, and the little ones will readily squeeze under fences (and also through them if the openings are large enough). The same is true for gates.

As with our chickens, I did my homework regarding goats and their needs before we got them. I was not totally inexperienced with goats, having had a couple of Toggenburgs during my college years. I knew how to milk a goat, and that they have minds of their own. But we were starting from scratch now and there was much I didn't know. I wanted to do things right.

Although we did not have a barn when we bought the place, we did have two old outbuildings. One of them seemed suitable for critters, and because it had two rooms, we thought we could use it for both goats and chickens. The larger room we divided into a chicken coop and a storage/milking room. The smaller one would make a nicely sized box stall, large enough for several goats.

What we didn't have was fencing. Our five acres was a little over half cleared, with two large open areas, and the house and outbuildings in between. Eventually we wanted to fence in both clearings, but we decided to start with the weediest, most overgrown one to fence for our first goats.

That fence became our first major outdoor project on the homestead. Modifying the shed was relatively simple, but we'd never put up fence before. We read about fence posts, gates, bracing systems, and fencing wire. We watched how-to videos. We researched materials that were available locally and how much they cost. Those of you who've read my first book or my blog know that we learned how to put up fence with a library book in hand, referring to it every step of the way. Dan also looked at livestock fences elsewhere, noting the types of posts used and the way bracing was done. We learned a lot from our research and from his observations. Dan would probably tell you he can now install fence in his sleep—not that he would want to.

Two types of posts are used for fencing: stout wooden ones for corners and bracing, and metal t-posts for line posts in-between. We were fortunate to find cedar posts for sale in the local paper for two dollars each, a real bargain. Eastern red cedar is very good for fence posts due to its resistance to rotting. The posts need to be stout (at least six-inch diameter) and about eight feet long for a four-foot tall fence. To install these a four-foot deep hole must be dug, either by hand with a post hole digger, or with an auger. Augers can be hand-held or PTO-powered farm tractor attachments. We did not have a tractor, so we bought a second-hand gas-powered two-person auger.

The gas-powered auger turned out to be more trouble than it was worth.

H-brace with horizontal cross member, cross wires, and twitches.

The videos made post hole digging look like a piece of cake, but then, they were digging in nice loamy soil with no rocks or tree roots. We have a thin loamy topsoil with a heavy clay subsoil. While there were mercifully few rocks, there were extensive tree roots along the edge of the woods. On occasion we were able to dig one of those piece-of-cake post holes, but too often we hit a tree root or a rare rock, which would cause the auger to buck and throw off its manhandlers. Sometimes it would drill round and round, seemingly going nowhere. In the end, Dan thought the manual post hole digger was quicker and more efficient, so we sold the auger for the same price we paid for it.

At the corners, gates, and long stretches, bracing is necessary. Braces reinforce the fence, and without them, the tension of the stretched fence will eventually pull down the corner posts. I've seen this before and it isn't pretty. Each corner requires three wood posts. A brace in the middle of a fence line requires two. Gates require two on either side. Concrete can be used in the post hole to add stability, or gravel and dirt can be tamped into the hole around the post. This works well for us since we have clay soil. Dan adds crushed rock to facilitate drainage, whether he uses concrete or packed soil.

The next step is adding a horizontal crossmember between the two posts. This arrangement forms the letter "H," which is why it is called an H-brace. In addition, a double strand of wire is run diagonally between the two posts and twisted together with a stick called a "twitch." These things tighten the brace and keep it sturdy.

T-posts require a tool to install too—a t-post driver. This is a heavy rascal requiring some height and strength to use, at least for the six-and-a-half-foot t-posts necessary for a four-foot tall fence. They need to be sunk about two feet into the ground. I gave it a try but couldn't manage it, so once we sold the auger, installing all the posts fell to Dan.

For t-posts, another very useful tool is a t-post puller. Eventually one will be needed and they make pulling t-posts out of the ground a breeze. Trying to dig and wrestle them out is not a breeze.

A t-post puller is a handy tool to have.

The wire fencing for goats should be the sturdiest one can afford. We looked at the options and compared them by price-per-foot. We needed close to 1000 feet for that first field, and with a limited budget, cost was a huge consideration for us. Field fencing was the most economical, about 50¢ per foot, but we dismissed it because I thought the openings were too large for goats. Woven goat fencing was available by special order, at about $1.00 per foot at that time. Cattle panels are very sturdy and used by some folks; they came to about $1.25 per foot. Welded wire was the cheapest at about 65¢ per foot. This was the option we chose because it was the only way we could afford all the fence we needed. I'd like to add that prices have changed since then; the welded wire is more like 70¢ per foot now, while the goat fencing is now locally available for about 75¢ per foot.

Some folks like to add an electrified hot wire on the inside of the fence. This is something to consider for a couple of reasons: one, it will prevent goats from leaning their full body weight into the fence as they walk along to scratch their sides; two, it can deter goats such as bucks with horns from trying to pull, push, or ram the fence when there are willing lady goats on the other side. I once met a woman who had only a single-strand electric fence for her goats. She'd had problems because tree branches often fell on it, so she turned it off. Her goats were none the wiser, however, and stayed inside their designated area.

We stretched our fence with a come-along and fence stretcher, then Dan stapled the wire to the wood posts, and I clipped it to the t-posts.

The come-along (manual winch) is attached to the fence stretcher to pull the fence as tight as possible. It is removed after the fencing is attached to the posts.

A fencing tool twists the clip around the wires to secure it to the t-post.

A small fencing tool and pair of pliers are used for this, using four or five clips per t-post.

Lastly we installed the gates. For our first gates we were less than precise, and they didn't fit perfectly between the gate posts. This was later remedied with a couple of nails and a 2x4. The bigger problem has been the unevenness of the land. Being in the foothills of the Southern Appalachians means the ground is rolling with little dips and ridges. Chickens, little goats, and little pigs take advantage of this and crawl under the gates to get from one area to another. Blocking these under-gate gaps with rocks and logs is our answer for that!

We started that first fencing project in September, but winter rains kept the ground too wet for installing posts. Gradually we got it all done and finally finished at the end of the following April. By May we were ready for our first goats.

Beginner Goats

I was excited about the prospect of getting goats. Why goats instead of a family cow? Dan and I talked about it, but since neither of us can digest cow's milk the discussion didn't seem worth pursuing. Plus we had a lot of brush needing to be eaten; a cow would do better on pasture with grass. Even if the first two considerations didn't count, a cow would be considerably more expensive to buy and would produce more milk than the two of us could use. So goats it was. But what kind? I contemplated this as I helped Dan level brace posts and clipped the welded wire to the t-posts. It was fun reading about the different breeds in library books, but when anyone asked what kind of goats we planned to get, my answer was always, "It depends on what's available when it's time to buy them."

Dan and I had different leanings when it came to goats. He tended to think "meat," whereas I tended to think "milk." Both of these were reasons we wanted goats, but, initially, the job to be done was tending to our badly overgrown field. One book suggested that dairy breeds are not the best choice for brush clearing, mainly due to the potential for udder injury. Meat breeds would be suitable for brush clearing, but meat goat breeders recommended supplementary feeding to keep the best weight on them. Scrub or Spanish goats would be best for the job, but I'd never seen one for sale around here. Another source suggested that it's best to get a combination of standard and dwarf goats, because they focus on different levels of brush clearing.

Shetland sheep were another possibility, because Shetlands are one of the few sheep breeds which love weedy browse. As a handspinner, there was a lot of appeal for me in this idea. I did some research and learned that Shetlands can gobble down kudzu and poison ivy along with the best of the scrub goats. Then I had a chance to talk with a gal in my weavers' guild who raises Shetlands. Her caution was that if the area had thorny briar plants, then I'd be continually "rescuing" the sheep when their fleece got caught in thorns and brambles. Unfortunately, that field was loaded with wild roses, saw briars, and sprawling blackberry bushes. Because of that, I thought it would be wiser to wait on the Shetlands, and ditto for fiber goats like Angoras.

I kept my eye on Craigslist and our weekly area sales paper long before we were ready to buy. It didn't make sense to set my heart on a goat breed that I would have to travel hundreds of miles to get. I gleaned several bits of information by regularly perusing these. At the time, the most widely available goats were various Pygmy types followed by Boer meat goats. There was an occasional Tennessee Fainting goat (Myotonic, also a meat breed) for sale or, more rarely a dairy breed such as LaMancha, Nubian, or Alpine. These were usually crossbred and usually bucks. Nearly everyone had a "billy" (male) they'd like you to buy. There was an occasional "nanny" (female) for sale, usually with her buck kid. On rare occasion wethers (neutered males) were offered and there were even a few Boer "steers" to be had. A good number of the goats for sale were "pet quality only." Prices ranged from $35 to $200 for mixed breed or unregistered animals.

Bucks are out for several reasons. Quite a number of years ago I'd had that Toggenburg doe and her rapidly growing buck kid. At the time they freely roamed the entire forty-acre farm and went where they pleased. They pretty much stayed on the farm unless I set out for town on my bicycle. Then they would try to follow me. I would have to sneak off when they weren't around. There was also competition between that buckling and me for his mother's milk. He began to challenge me so that, eventually, it was no fun having goats. After that experience I did not want bucks. I figured we could keep a couple of does and get buck services elsewhere. Except for wanting that milk for yogurt and cheese, I wasn't interested in becoming a breeder, although Dan talked about the possibility of someday raising and selling meat goats. For me, there was a vague possibility of someday keeping and raising a heritage breed of goat, but with only one fenced area for goats I wouldn't have to think about getting bucks for a long time.

Out of curiosity I did some internet searching for local goat breeders. I found I could get purebred Nigerian Dwarfs, Toggenburgs, Boers, or Kikos (another meat breed) with or without papers. These were pricier—$350 and up for a registered animal. There are advantages to buying from a reputable breeder, particularly being able to purchase animals from someone with experience who can provide documented health and vaccination records, and whose reputation depends upon the quality of their animals and customer satisfaction.

With all of this, I was beginning to narrow down the realistic possibilities. I would need a minimum of two goats because they are herd animals. A lone goat would be miserable. Four would probably be a good number for us, considering how much there was to eat in that overgrown field. It was approximately one acre and could probably support more initially, but I figured we'd have our hands full if we did get four.

One possibility would be to go with a few meat goats and one dairy doe. That would make us both happy. For dairy, breed was a consideration because Dan didn't care for the look of either LaManchas (earless), or Nubians (droopy ears and Roman noses). He liked the look of Alpines while I leaned toward Toggenburgs, because that's what I used to have.

Toggenburgs run on the smaller side of standard-sized dairy goats. They have a long milking season but their milk is lower in butterfat (cream) than the popular Nubians and Nigerian Dwarfs. The downside to the lower-fat milk is that some folks don't think it tastes good. I thought Toggenburg milk tasted just fine, but I wanted Dan's first taste of goat milk to be good. I wanted him to like goat milk and decided against the Toggs.

In the end it all boiled down to availability and price. Shortly after our fence was finished I bought our first two goats. They were a mother and daughter, mostly Boer with a touch of Nubian somewhere along the line, or so I was told. I say that because the fellow who sold them to me changed his story several times when I went to see them. I never responded to his comments about breed and he finally looked at me and said, "I don't know what you're looking for!" Very revealing, but the breed didn't matter as much as purpose and price. He was asking $100 for the two of them and I thought they would be perfect for that overgrown field.

I brought them home late Monday afternoon, in a big dog carrier in the back of my Jeep Cherokee Sport. I named the mother Abigail

Bathsheba (aka CryBaby) and Abigail, first goats on the homestead.

and the doe kid Bathsheba because of the movie *Seven Brides for Seven Brothers*. Abigail and Bathsheba weren't characters in the movie, but the hero's mother had named her seven sons alphabetically with names from the Bible. For some reason I liked that and wanted to do the same with my goats.

Abigail and Bathsheba spent the rest of the day and night in their new home, the right side of our old converted shed, with the chickens in the left side. The chickens had never seen goats before and were very curious. Our cat Rascal had never seen them before either and was very worried. The goats were used to chickens and cats and took it all in stride. I would like to say it was without mishap, but in the morning I found Abigail with her horns stuck in the nifty hay feeder we'd made from a cattle panel. There was minor panic on my part, but she was freed without injury or protest.

Having been part of a large herd and not handled very much, neither was particularly tame, nor particularly wild. I wasn't interested in making pets of them, but I did need them to be tame enough to handle as needed. After a couple of days they would come when they saw me, but kept a respectful distance. Abigail would back away if I put out my hand, and Bathsheba would follow suit. Within those first few days it became apparent that little Bathsheba's name didn't suit her very well. She quickly earned the name "CryBaby."

After we got Surprise, we had to divide the goat stall into two.

About a month later, I bought my first dairy goat. She was a registered Nubian, something I hadn't seen before on Craigslist, so that was exciting. The name on her registration papers was Precious Memory Farm Surprise. She was eighteen months old and had possibly been bred. I wasn't sure if I was ready for kidding yet, but at the time I hadn't seen many Nubian does for sale and didn't want to pass her up.

I had already decided that I wanted a dairy goat in addition to brush goats. I could have bought a mixed dairy breed for less, but other ideas were rolling around in my head. What changed my mind?

First, the milk itself. As I researched dairy breeds I read over and over that folks generally prefer Nubian milk. That was important to me because I needed Dan to like goat milk. Not that we drink milk, but we eat a lot of yogurt, kefir, and cheese. If we were going to keep goats he needed to like what they produced.

But why the expense of a registered Nubian doe? I surprised myself with that decision as well, and this is the second reason I bought her: as I researched goats and goat breeds, I had learned about a breed that I thought might be perfect for our homestead: the Kinder.

Kinders are mid-sized, dual purpose goats and seemed perfect for homesteaders. As a cross between Nubians and Pygmies, they inherit the very best of both breeds: the good milk production of a dairy breed, and the heavier muscling of a meat breed. There are two ways to obtain Kinders: by either purchasing them or by purchasing a "Kinder Starter Kit." This consists of at least one registered Pygmy buck and one registered Nubian doe.

Kinders were another breed not available in my area. That left me two choices if I wanted them. I could either transport Kinder goats hundreds of miles from another state, or, I could start my own herd. This would require a registered Nubian doe to eventually be bred to a registered Pygmy buck. When Surprise unexpectedly showed up on Craigslist, I didn't hesitate. She was my first step in that direction and the beginning of my very own Kinder Starter Kit.

With the addition of Surprise, I was introduced to the social structure of goats. Surprise absolutely did not want to be with Abigail and CryBaby, and made a beeline out the gate every time it was open. Abigail didn't want her around either. She considered the stall and hay feeder to be theirs, i.e. hers and CryBaby's. She was forever chasing Surprise away from the food and the barn. Since she was horned and Surprise wasn't, this was a concern.

At first I assumed this had to do with maternal instincts, but even after Surprise had been around long enough to obviously be no threat, Abigail was merciless in bullying Surprise. CryBaby, on the other hand, loved Surprise and wanted to be best friends.

Trying to figure out what to do, I searched the internet. Someone on a goat forum suggested that the goatkeeper should make it clear he or she is the alpha "goat." I decided to give this a try. Every time Abigail would make a threatening move toward Surprise, I would lower my head and lunge toward Abigail as if to say Surprise was off limits. Abigail would back off. We did this several times over several days and I began to think everything would be okay. Well, it was okay as long as I was around, but when my back was turned it was goat business as usual. I tolerated it until Abigail cornered Surprise in the shed one day, ramming her until her shoulder was bleeding. That was it; Abigail was a goner and I sold her.

CryBaby frantically searched for her mother (in between mouthfuls of grass and browse) for about a day, and then we got our llama, Charlie. Never mind mom, what in the world was that huge fuzzy thing in our barn?! Everything else was forgotten in the presence of this strange furry creature. CryBaby attached herself to Surprise pretty quickly, adjustments were made all around, and things in the barnyard seemed to settle down.

It turned out that Surprise was not pregnant, which was okay with me. I had a little bit of experience now tucked under my belt, and I felt like I was entering a new phase of goat keeping. That was a good feeling.

SURPRISE

"What are you going to do about the Princess?" Dan asked. He was referring to Surprise. She and Alphie, her three-day-old buck kid, had been residing in the kidding stall. It was time for them to be moved out. I had another doe due any minute and I needed that stall for the impending birth. Dan and I both knew Surprise wouldn't like that.

As my first serious goat purchase, I confess I pampered Surprise considerably in the beginning. She was an investment in the future of our homestead. Besides, with the way Abigail bullied her, I felt sorry for her. Little did I know she would be my all-time, most annoying goat ever.

Surprise was fickle. That first summer my critter population changed quite a bit. I bought Abigail and Crybaby, then Surprise, then sold Abigail, bought Charlie Llama, and then bought Jasmine. Jasmine was my second registered Nubian doe. These four, Surprise, CryBaby, Jasmine, and Charlie, shared the goat shed and paired off in a buddy system: Surprise with CryBaby and Jasmine with Charlie. That seemed perfectly acceptable to everyone until I lost Charlie unexpectedly.

Surprise dropped CryBaby like a hot potato and immediately attached herself to Jasmine. Poor CryBaby was the odd goat out.

Although Surprise mellowed out over the years, she has probably been the loudest, most vocal doe I've ever had. This is characteristic of Nubians, but she was nonstop. The only time she seemed to be quiet was when she was trying to undo the chain on one of the gates. Most of our gates have slide bolt latches, which I suppose she figured out that she couldn't figure out. Every morning she would head for that chained gate and work on trying to undo that chain.

Surprise refused to acknowledge the routine. She'd holler and shove and push, and generally get in the way at feeding time. That is, unless she spotted the gate open and then, zoom! Out she'd go. She had a sixth sense when it came to anything I didn't want her to do, or to get into. That would be the very thing she'd immediately home in on and go for. If I had a sheet of plastic covering something to protect it from rain, there she'd be, yanking and tugging until she tore it and pulled it off. She could be persistent about this, too. I once had a pile of straw and manure I wanted to save for sheet composting the pasture we planned to reseed in the fall. I made the mistake of thinking I could leave it next to the goats' shed and simply cover it to keep the chickens out. Surprise would not leave that pile alone. The more bricks and rocks I put on top of the plastic to hold it down, the more determined she was to get to it.

This little "game" of ours became increasingly annoying until I finally decided to go full guns. One day, while the girls were grazing elsewhere, I piled bricks, cinder blocks, short lengths of metal roofing, my compost sifter, a huge plastic tray, and the overturned wheelbarrow on top of the sheet of plastic covering that pile. As an additional deterrent, I dragged the garden hose over to the gate, setting it just out of sight of the goats.

When the girls returned to the goat shed, they immediately checked out the pile. Jasmine lost interest pretty quickly, and apparently, Surprise did too. But had she really? Or was it because I was standing there watching? Sure enough, when she thought I wasn't looking, she sneaked around to the other side of the pile and started tugging on the plastic. I turned on the hose and let her have it! Needless to say she had no success after several rounds and gave up. Score one for me, finally! This became part of our daily routine and the straw pile survived, although the plastic was somewhat raggedy by the end.

Surprise was sneaky in other ways too—on the milking stand, for example. In addition to the stand, the milking room contains shelves for various goat supplies and large plastic trash cans for storing feed.

My efforts to deter Surprise did more to rouse curiosity than to keep her away.

Goats hate water, so a squirt from the garden hose was the most effective way to keep Surprise from tearing into the pile. The problem was that I couldn't stand around manning the hose all day.

Sometimes after getting her on the stand I would forget to secure the head gate. When latched, the head gate keeps the goat from backing out and jumping off the stand. If I forgot to secure Surprise's head, she would wait until I stepped out of the room to check on a crying kid or squawking chicken. As soon as I was out of sight she was off the stand and over to the feed containers. She would flip the lid off and help herself to the contents, even though that very same goat grain remained untouched in her feeder. She got pretty good at lid flipping too, although admitting that is a confession as to how many times I let her get away with it.

Our biggest battle, however, was over the kidding stall. When we expected our first crop of kids, we divided the 10-foot-by-14-foot goat shed into two stalls. This gives me a way to separate an expectant doe and give her privacy for birthing. It also keeps bigger goats out until the newborns are steady on their feet. Surprise was the very first doe to occupy the kidding stall. For some reason, she liked having that stall all to herself and claimed it as her own from that day onward. Even Jasmine, who could pull rank over Surprise when it came to treats and food, respected Surprise's territorial claim to this stall. Anytime I moved Surprise out and another doe in, Surprise was not pleased.

That kidding stall was the object of the biggest battle of wills between Surprise and me. This points to something I have learned about goats over the years. They absolutely do not see humans as the superior or dominant species. Humans seem to think that since they "own" the goats, the goats should do things the way humans want. Goats don't see it that way. My Toggenburg buckling and Abigail are two examples. I've had other goats (bucks mostly) who would challenge me to an occasional standoff to determine our respective places in the homestead's human/goat social order.

Surprise and Alphie. He turned out to be a "chip off the old block."

Most goats don't care as long as you feed them, but some will gladly dominate humans. These require the human to prove that he or she is worthy of a goat's respect. If Surprise couldn't have her stall, then she absolutely was not going to cooperate with me. The battle lines were drawn. Her arena of choice? The milking stand.

There are several ways to manage goat kids and milk. One is to separate the kids from their mother at birth and bottle feed them. There are several reasons for this method, such as a commercial dairy wanting to sell all the milk, breeders keeping milk production records, or managing triplets or quadruplets to ensure all the kids are getting their fair share of milk. Bottle feeding also becomes necessary if a doe rejects a kid, or if she is positive for CAE (Caprine Arthritis Encephalitis). This dreaded, incurable goat disease is passed on through colostrum, milk, or other body fluids. It does not affect humans but it is deadly for goats. The virus is killed by pasteurizing the milk, so bottle feeding is necessary. Some breeders practice a CAE prevention program, whereby they routinely feed only pasteurized milk to their kids, regardless of the doe's CAE status.

Bottle feeding is a lot of work, however. The simpler and preferred method of many goat keepers is to share the milk with the kids. This is done by separating the kids from their mothers at night once they are old enough to nibble hay. The goatherd gets the morning milking, and the kids get it for the rest of the day. The beauty of this method is that morning milking can be whatever time is convenient for the person doing the milking. The kids are separated about 12 hours prior and that's it. This is the method I use. In the case of Surprise, who produces more milk than even twins can initially consume, I start milking her from day one. I've been able to collect and freeze colostrum for future emergency use, plus we get an earlier start on fresh goat milk.*

Now Surprise was mad at me for giving away "her" stall. To prove it, she refused to get on the milking stand. At first I thought this was because I'd moved it and she didn't like the new location. Like all creatures, goats like things to be familiar. They like their routine. I tried to talk reassuringly and coax her on with food. She refused. I tried to nudge her onto it. She wouldn't budge. I tried to drag her up and onto that stand but she planted her feet and resisted with her full weight. I sighed and moved her feeder to the ground, thinking I'd milk her where she stood. I'd done this in the past with no problems so I didn't anticipate any now. Oh, she was hungry and wanted her grain all right, but she was not going to let me milk her. She sidestepped. I moved her

*The obvious recommendation for this method is to purchase goats that either test negative for CAE or come from a CAE-negative herd.

closer to the wall. She kept on eating and I managed to get her udder washed and the bucket in place, making sure I kept one hand on it in case she kicked. I started milking with the other hand but nothing happened. She refused to let down her milk.

The letting down of milk is something over which goats have control. Kids will bump their mother's udder with their heads to signal her to let down her milk (or more milk). For the human milker, massaging the udder with a warm washcloth accomplishes the same thing. With the morning-only milking method, a doe can learn to hold back some milk for her kids. If she is nervous she may be reluctant to let down her milk. Or, as in Surprise's case, if she is holding a grudge against the goat keeper she may not let down her milk. I managed to massage and coax some milk out, but eventually she managed to step in the bucket with her hind foot.

Milk that has had a goat foot in it goes to the chickens, and the chickens got the milk for the next several days. I knew I somehow had to get Surprise onto that stand where I could restrain her head and perhaps her hind feet. The next day I tried forcing her onto the milk stand by pulling on her collar. She refused. I insisted. She was adamant. I could have used another pair of human hands but was on my own, so I had to come up with my own solution. I snapped a leash onto her collar and ran it over the feeder, through the head gate of the milking stand, and then under the milking stand where I could grab onto it from the back. I figured this way, I could pull on the leash and push her rear end onto the milking stand at the same time. Well, it worked and she jumped onto the stand. I secured her head in the head gate and she began to eat. Whew.

No sooner did I get the milking bucket positioned and put a hand on her teat, then she started to buck and kick. She not only knocked the bucket off the stand, but I lost my balance and landed hard on my backside as well. By that time I was pretty fed up. I said, fine, and put her back in with the other goats, unmilked and without the rest of her grain. She wasn't expecting that! Even with my less than successful attempts of the previous days she had gotten her grain. This morning she'd have to make do with hay and browse only.

Even with Alphie getting his full share of milk, Surprise's udder was extremely full and heavy the next morning. She waddled to the stand and I hoped that in her discomfort, she would want relief enough to cooperate with me. She did, sort of. She climbed onto the stand but front feet only. Her hind feet remained on the ground. I could work with that and proceeded to wash her udder and milk her. She didn't kick once but ate all her grain ration while I got a full bucket of milk.

That was how we reached our "agreement," the terms of which, were in force for as long as I had her. She would come to the stand and permit me to milk her, but she refused to jump all the way up, preferring a front-feet-only stance. Sometimes she would forget herself in her enthusiasm for her feed and jump up all the way. Then she would remember and try to jump back down unless I secured the head gate quickly enough. But there was no more balking and kicking. She continued to think that stall was hers, but at least I got the milk.

JASMINE

If Surprise could be considered my most annoying goat, Jasmine was my most frustrating. This had nothing to do with personality; she was friendly, easy-going, and even-tempered. She had a way of cocking her head sweetly when she looked at you that made you want to pet her. She never butted or bullied the other goats, and was patient with all their kids. Everybody loved Jasmine.

Jasmine started her residence here like the others, in the goat shed with the gate closed until introductions could be made. I don't believe she'd ever seen chickens before, so she was a bit cautious about them. She'd definitely never seen a llama, and she wasn't all that thrilled about her new caprine companions. She just wanted to go home.

After the others lost interest and wandered off to browse, I let Jasmine out. She stuck to me like Velcro. She kept her face pressed to my backside as I walked her out into the field. I had to sneak off by taking gradual backward steps because she hollered every time she thought I was leaving her out there alone with all those strangers (melodrama is a Nubian specialty). Happily, she made a smooth social transition with no standoffs or sparring for position. Of us two humans, Dan was her favorite. She liked me well enough and always came to greet me, but she'd abandon me any time she saw Dan.

Like Surprise, she was a registered Nubian, the second in my Kinder Starter Kit. She was quite a bit bigger than Surprise but not as demanding. Most of the time Jasmine let Surprise have her way. Until it came to food. Then Jasmine had dibs. If I brought the goats leftovers from garden harvest or canning, Jasmine would get first pickings, which never failed to trigger a temper tantrum from Surprise. Surprise would stamp her feet and holler, but Jasmine would block her way. Surprise would take off in a huff, running, bucking, and bellyaching. She would get about halfway to the goat shed, realize she was going in the wrong direction (as in away from the food), then turn around and go charging back again, complaining the entire time. In the meantime Jasmine would calmly finish off the treats. Surprise always forgave her, though, and if goats could ever be considered best friends, Jasmine and Surprise were.

Jasmine was frustrating because of health issues. When I first went to look at her, the seller told me he was drying her off (stopping milk production) because her udder was producing milk on only one side. He went on to explain that the condition had been treated and once dried up would be fine. At her next freshening she should produce on both sides again. I took his word for it and bought her. Since this was my first experience with drying off, however, I did a little research.

There are two methods for drying off a doe: gradually or cold turkey. The gradual method is what was being used on Jasmine by slowly increasing intervals between milkings. This would signal her body that less and less milk is needed so that she will produce less. By the time I brought her home she was being milked once every three days or so. The cold turkey method is to simply stop milking. According to Dr. Susan Kerr of Washington State University, the idea was to wait until the doe was producing less than three pounds of milk per day, feed her a high fiber, low grain diet for about two weeks, then stop milking until the next time the doe kidded.[1]

I really had problems with this idea. For one thing, I'd experienced milk engorgement when I was breastfeeding and knew how painful it can be. For another, left to their own means, mama goats slowly wean their kids, letting them nurse less and less until about five months of age or so. Why would I want to do it differently? Still, Jasmine was down to being milked every three or four days, so I figured I could just stop. I didn't think anything of it for the next several months.

I became very much alarmed then, when I noticed scabs on Jasmine's bad teat. I took a closer look and saw that the right side of her udder was hard, had scabs, and one weeping sore. I hadn't paid a whole lot of attention to her udder since I stopped milking her in September. I felt very much ashamed that I hadn't noticed this immediately. The teat wasn't hot to the touch which would have been indicative of mastitis, so I didn't know what to think. The next morning I took her to the vet.

He said it was definitely infected, and prescribed penicillin injections every other day for a total of four doses. The bad news was that there was some necrosis of the tissue. He said this would slough off, but that the teat would probably seal itself off in the healing process. If that happened, it would mean she couldn't be milked on that side. He went so far as to suggest that if I was planning to cull any goats, she would be a candidate.

Well, I was neither ready to cull her nor give up hope, and felt that I had to do something. The only thing I knew to do besides the prescription medication was to try natural remedies. I started by giving her garlic as a natural antibiotic. Fresh garlic contains sulfur compounds, specifically allicin, alliin, and ajoene. These are what make it a powerful anti-bacterial, anti-viral, and anti-fungal agent.[2]

I've changed the way I've given garlic to my goats over the years. Poor Jasmine got it liquefied with water in the blender—a couple of cloves per 1/4 cup water. This was administered orally with a dosing syringe, about 10 cc per dose. She hated it but got a nice chunk of apple as a chaser and reward. Later I learned I could add blackstrap molasses. Once the goats got a taste of the blackstrap, they would come running anytime they saw me with the syringe. Even easier is to feed the garlic raw by adding whole, unpeeled cloves to their feed. If they need it, they'll eat it. I offer it at the morning milking and have never had milk tainted with a garlic taste.

Another thing I did for Jasmine was to make an herbal salve based on Dr. John Christopher's "Bone, Flesh & Cartilage" formula.[3] It contains white oak bark, comfrey, mullein, marshmallow root, black walnut hulls, gravel root, wormwood, skullcap, and lobelia. It definitely wouldn't hurt and I had every reason to hope it would help. I massaged her udder with it several times a day. Massage is beneficial because it helps stimulate circulation. Good circulation helps bring oxygen and nutrients to that area of the body and removes cell waste at the same time.

Slowly, gradually, Jasmine's udder healed, and I felt triumphant that the teat did not slough off. What remained to be seen, was whether or not the teat would function normally again. Jasmine had her turn visiting our buck, so I hoped I would learn the answer to that question soon.

Herbal salves can be comforting as well as healing, and they are easy to make. See "Resources" for where to find how-to instructions.

I keep potential due dates for my goats marked on the calendar. When they come to within a week of those, I put them in the kidding stall at night. The closer the date gets, the more frequently I check on them. It was a little after 3:30 a.m. one May morning when I checked on Jasmine and discovered that she was in the second stage of labor.

Everything seemed to be going normally at first, but after a while I began to worry that it was taking too long. Once the doe reaches the pushing stage, the birth should be within the next 30 minutes. It had been 45 minutes and nothing else seemed to be happening. This was only my third kidding, so I reviewed my books, carefully washed my hands and arms, and went in to investigate what the problem was. I was supposed to feel for legs but couldn't find any. What I did find was a rear end and a tail. This is called a frank breech and the doe requires help or both she and the kid will die. My job was to try to find the hind legs and guide them feet first into the birth canal. This I was able to do, but even then, labor seemed to take forever. Jasmine was exhausted and the doeling was stillborn; the cause of death was likely compression of the umbilical cord, which cut off oxygen.

The next morning I took her to the vet. I was uncertain about the placenta, plus the right side of her udder (the "bad" side) was hard. I could not express milk from it, so I was very concerned about pressure buildup as she continued to produce milk. It would be painful for her, and my vet once told me they'd seen udders rupture from this. Because of all of Jasmine's problems, I decided to take her to someone who specializes in larger animals and ruminants. My local vet is excellent for routine things, but for this, I wanted to take her to someone who was best set up to help her. It meant a long drive to another part of the state, but this was our best option.

I had started Jasmine on penicillin injections at home, but her temperature at the vet's was 104.4° (40.2°C. Normal for goats is usually considered to be 102° to 103°F or 38.8° to 39.4°C). The vet examined her and said she was developing a uterine infection, which is not uncommon under the circumstances. The vet flushed her out, and gave her several injections: Nuflor for the infection, Banamine for the fever and swelling, and Lutalyse to help her expel anything that might be left in the uterus. I was also given two doses of Pirsue for the mastitis. We made another trip for a second uterine flush and more injections.

Jasmine's appetite remained fair, so at least she got fluids and nutrition. I added probiotics to her diet because of the antibiotic. I added blackstrap molasses to her drinking water to make sure she took

Jasmine with Surprise's twin bucklings. Jasmine was very tolerant of the other doe's kids. In fact, I never saw her butt anybody.

lots of fluids. Emotionally, she was loud and clingy, hollering pathetically unless I was with her. All she seemed to want was body contact. She had a mellow personality for a Nubian, so this was very unlike her.

My high hopes for a healed udder were shot down by the reality of our situation. The orifice of the teat was sealed off so that she still could not be milked on that side. I decided I had no option but to dry her up as quickly as possible. Toward that end, I searched the internet for herbs to dry up milk supply. There was a lot of information for humans, and I figured they would work for Jasmine as well. One herb reputed to help is sage, which I grow in my herb garden. I started adding that to her feed.

As I continued my research, I ran across an article entitled "Herbs, Milking and Mastitis, Five Secrets."[4] I had already given Jasmine the vet's prescription antibiotics for mastitis, which hadn't helped, so I was willing to give the herbs a try.

Three of the four herbs mentioned to treat mastitis were ones that I had growing on the land: poke root (traditionally used to treat problems with the lymphatic system, skin problems, tumors of the mammary glands, and mastitis), comfrey leaves (used for skin conditions, sprains, bruises, ulcers, burns, and swelling), and calendula flowers (applied to cuts, scrapes, rashes, and inflammation). The one recommended herb I didn't have was red clover, traditionally used for tumors, skin conditions, and breast discomfort. I just used what I had.

It was easiest for me to make a salve, so I pulverized these in the blender with vegetable oil. I admit I didn't measure. I just dug a big chunk of poke root, grabbed a handful of comfrey and another of calendula flowers. Once I had a slimy green goop, I added melted beeswax, and let it cool. As the beeswax cools, it gives the salve a less liquid consistency. I ended up with a pint of salve, which I massaged onto her udder three times a day.

I noticed no difference for the first several days, but gradually the bad side of her udder was no longer swelling with milk. I was greatly relieved at this. I stopped giving her sage, since she didn't care for it anyway, and continued milking out the good side twice a day. Instead of the sage, I offered her fresh comfrey leaves daily, which she ate with relish. She loved having her udder massaged with the salve. She'd bow out her legs to let me rub and massage for as long as I wanted. Within a few days the mass in her udder was noticeably smaller. In three weeks it went from a rock hard melon-sized mass, to a tangerine-sized softer (but still palpable) lump.

The biggest question on my mind was whether or not I could (and should) breed Jasmine again. The reason I'd bought her was because I needed a registered Nubian doe as part of my plan to start my own Kinder herd. I discussed it with the vet, who seemed to think that Jasmine would do fine, especially if she was bred to a Pygmy. The kids would be smaller and the birthing should theoretically be easier. I decided to give it a try. We let her have her turn with Gruffy, our Pygmy buck, and waited.

While we waited, Jasmine developed hoof rot. Hoof rot is bacterial in cause and usually transmitted in warm damp conditions such as mud. According to Pat Coleby, goats with copper and sulfur deficiencies are especially susceptible. I followed her advice and soaked the foot in a copper sulfate solution.[5] It was a hassle to get Jasmine to stand there with her foot in the solution because goats hate getting wet. After the soak I was supposed to sprinkle the copper powder on her hoof and bandage it. This was less successful because Jasmine managed to work the bandage off. Still, we managed and eventually the problem was resolved.

By this time I was getting weary of all the special treatments Jasmine had required. The time it was taking was causing me to wonder if it was worth it. I just wanted Kinders. What I had gotten so far was a parade of problems that took time and resources away from other needful things.

While waiting on our due dates for both Jasmine and Surprise, I decided that once I got my Kinders I would sell Jasmine. I knew it wouldn't be easy considering her udder condition, but I also hoped someone else would have the time and resources to care for her better than I could. It was a relief to come to this decision, and it lifted an emotional weight from my heart.

One day after a brief thunderstorm I walked to the goat shed to discover Jasmine standing there, dangling her left front foot. She would put no weight on it and I knew immediately that something was wrong. Initially we thought it was dislocated. We surmised that she had made a dash back to the shed when the rain started and had slipped as she rounded the corner. We took her to the vet. His diagnosis wasn't good; she had broken her shoulder. The prognosis was worse. The fracture couldn't be set like a leg fracture could; she would likely be severely lame for the rest of her life. He recommended putting her down. We weren't ready to do that, especially since she might be pregnant. We had a pregnancy test done and took her home to await the results.

Now that she couldn't get around, her level of care was higher than ever before. I could give her the prescribed pain medication, even herbs recommended for broken bones. However, if the bones healed incorrectly—well, I didn't want to think about that.

When the pregnancy results came back negative we knew we had to make a decision. It had been several weeks of daily Jasmine chores: cleaning out her stall, adding fresh straw, and bringing her laundry baskets of fresh forage. I was getting tired of it and so was she. She eventually quit getting up and I knew it was cruel to allow her to go on like that. We discussed the options. Everything we had said about keeping an emotional distance and not getting too attached to our critters became larger than life. Even though we had planned to raise goats for meat, we'd never thought of Jasmine that way because she was a dairy breed. I had always assumed that after she gave us several years of Kinder kids, I'd sell her. Now our only choices were to shoot and bury her, or take her to a meat processor.

In the end we took her and our wether to the processor. Intellectually it was the best decision; even in death these goats would still have purpose in nourishing us. Emotionally it was harder. It was hard not to wonder if we could have prevented it or if we could have done something more. The fact of the matter is that sometimes things happen even when we're doing our best. Sometimes there is personal responsibility to be taken, other times it is no one's fault. Some things cannot be fixed. Some things can only be accepted. Homesteading with animals was really bringing this to the forefront.

We were exceedingly thankful for that meat in our freezer. To not have made that choice would have seemed like a waste of her life. It was our first chevon and it was good. Jasmine taught me a lot, and for that I will always be thankful.

LITTLE CHIPPER & OLD McGRUFF

Wanting something and making it happen aren't necessarily the same thing. What I was wanting was Kinder goats. Making that happen was proving to be a bigger challenge than I had anticipated. Registered Nubian does had been easy enough to find, but finding registered Pygmy bucks was something else again.

Nubian goats are seasonal breeders. This means they come into heat in the autumn, when daylight hours are beginning to decrease. Pygmies and Kinders, on the other hand, are aseasonal breeders; they will come into heat throughout the year. By the end of summer I had my two Nubian does, but with fencing for the bucks still in progress and no Pygmy buck, I needed to come up with a plan. The last thing I wanted was to wait another year.

PLAN A. Since the fence wasn't done, the best thing to do would be to find a registered Pygmy buck close to home and pay a stud fee to have my girls serviced. The problem with this plan was that no one around here kept registered Pygmies. So...

PLAN B. Find a registered Pygmy buck within no more than half a day's drive and pay a stud fee to have my girls serviced. This one seemed more hopeful. I found a breeder about 150 miles from here who offered stud services. The stipulations were that each doe required a veterinary certificate dated within 30 days of the breeding, to certify they were disease-free. The breeding would be "by hand," meaning the doe is introduced to the buck and removed once the deed is done. The plan seemed less appealing when I realized that it would mean a 300-mile round trip per doe, unless they both came into heat at the same time. The biggest problem was that goat breeding was new to me and I really couldn't tell when they were in heat. It's not as obvious as say, a dog or a cat (although I later learned they can be given hormone injections to make them ovulate). That consideration, on top of stud fees, testing fees, and travel expenses, made me reconsider. Wouldn't it be cheaper and easier to buy my own buck? So...

PLAN C. Blitz the fence and work on it 24/7 to get it up. Then find a registered Pygmy buck to buy. This would have been the most preferable option, but had the same problem as Plans A and B; while there were lots of Pygmies advertised for sale, no one seemed to have registered stock. The few registered bucks that I did find were not only far away, but also only about two months old. While they might be sexually active at that age, the prospect of getting one of these little guys to successfully breed full-size, grown-up does seemed slim to none. Options seemed to be running out. So...

PLAN D. With Dan being an over-the-road truck driver, I wondered if I could find a registered Pygmy buck near where he made his runs. There were lots of Pygmy breeders in Ohio, for example. I could make the deal long distance and he could do the transaction and bring a buck home from wherever. This plan wasn't very practical, however, as he had no way to actually transport it—no container or a large-enough dog carrier, and no place to put one. Besides, if he could do that, why not just bring home some Kinders? So...

PLAN E. Find a dairy buck of any breed for that year, breed my does, and worry about a registered Pygmy buck in the spring. This plan was successful. I bought a young, mostly Nubian with a little Alpine mix buckling. He was about seven or eight months old. I named him Petey and put him in with the girls. Five months later my first kids were born. I didn't have my Kinders, but I had milk!

That spring I finally found two registered Pygmy bucks for our newly-fenced buck pasture: four-year-old Sugarland Farm Rescue Me and two-month-old Tarheel Acres Chipper. Rescue Me had a gruffy, deep-throated bleat which made me think of "Three Billy Goats Gruff." We started referring to him as "Ol' McGruff," which quickly became "Gruffy."

The Pygmies made themselves right at home. The only problem was that Chipper, being a baby, didn't want to be alone. Of course he

Chipper was shy for the longest time.

missed his mom, and while Gruffy wasn't exactly the ideal substitute, at least he was better than nobody. Except Gruffy didn't know he was supposed to be Chipper's buddy. Poor little Chipper was so short that he couldn't see over the long pasture grass, so he would bleat and bleat for Gruffy. Gruffy paid no mind. I would have to go out to find Gruffy and stand by him while I called Chipper. Chipper would come running. He remained pretty suspicious of me, so he would make a wide detour around me to run behind Gruffy to hide. But at least he was no longer lonely and he quit hollering for a while.

Chipper was the first kid I lost in what became a series of mysterious kid deaths over the years. Even the vet was puzzled when I brought Chipper in. There was no diarrhea, no fever, no heavy parasite load. He just seemed to give up and then suddenly, death. I was devastated. How could I have let that happen? I grieved the loss of a little buckling I had both grown to like, and who was an important part of my Kinder Starter Kit. Even though he was too little for breeding that year, he would have certainly provided genetic diversity in the years to come.

What could I do but carry on? Gruffy was more than willing to do his part, and the girls seemed interested too, until it came to the actual deed itself.

The size difference between the two breeds was one problem, although Kinder breeders find ways around this, such as a bale of straw for the buck to stand on. We tried a small ridge, even made a small platform for Gruffy, but neither Jasmine nor Surprise would stand still, especially if we tried to hold them in place. In the end I left Gruffy in with the girls for several months, marking sighted breeding attempts on my calendar. As the due dates approached, I prepared the kidding stall and waited.

The only true ways to confirm pregnancy in goats is either by blood test or ultrasound. A huge belly isn't an accurate indicator because it might just be a full rumen. Most goat owners can tell you about fat goats they've had that were not with kid, and also surprise kiddings from goats that didn't look the least bit pregnant. Of course the doe will stop going into heat if she is pregnant, but hormone changes can also change her behavior in ways that mimic a heat cycle. Of blood testing, one can have a vet do it, or learn to draw blood oneself and send it to a lab to be tested. I did neither of these things. I simply waited and hoped.

It was during this time that Jasmine fell and broke her shoulder. When her pregnancy test turned out negative, I put all my hopes in Surprise. Every due date came and went until there were no potential dates left and no kids either. What a disappointment.

The following year I gave Gruffy several more tries with Surprise, but when I had doubts about success I faltered in my resolve to have Kinders. I thought about selling Gruffy, but by that time he had became a homestead favorite. With his big brown eyes, squatty body, and low mellow bleat, he was impossible not to like. He was always willing to come greet whomever entered his pasture and he loved to be scratched and petted. I could see why people loved Pygmy goats as pets. They have great personalities. So we kept him in spite of his not earning his keep. His keep was actually nominal, since he kept himself fat on browse and hay. I lamented that I wasn't able to pass that quality on to Kinder offspring, but had to accept that this was the way things were.

THE STORM

The summer was a hot one, not untypical for our part of the country. Our temperatures stayed in the upper 90°sF (30°sC) throughout July and August, occasionally topping 100° (37.7°). When it gets that hot, the ground is always thirsty. Summer's pop-up thunderstorms are always welcome, but the water evaporates right out of the ground shortly afterward. After four weeks of no rain, we finally got a real live, bona fide Texas-style gully washer one Friday evening. We were eating dinner when we realized it had started to rain. Not a gentle rain, but a pounding rain. My first thought was of Surprise's twin bucklings. They recently had been weaned and separated from their mother. Rather than spend the nights with the older bucks, they'd preferred to stay outside, in the corner of the buck pasture as close to their mother as they could get. They had no protection from the driving, pelting rain. I needed to get them to the buck barn as quickly as possible.

I remembered that my rain poncho was in the tool shed (darn) so I took off running without it. I found the little guys huddled under a pecan tree, crying their little hearts out. I knew that if I carried the younger one, the older one would follow.

It was raining so hard that I couldn't see. My long denim skirt was soaked and heavy, pulling at my legs as I tried to run toward the billy barn with one little goat in my arms and calling to the second one to follow. The wind was blowing so hard that I didn't even know if he could hear me, but I could hear him bleating from behind. I ran the best I could, but for some reason (possibly the atmospheric pressure) I couldn't breathe. I couldn't get air into my lungs. I had to will myself toward the buck barn, gasping as I went. Branches crashed around us, but thankfully we made it—soaking wet, but unharmed.

In the meantime, Dan had taken off to check on the does and the chickens. He showed up within seconds after I got to the buck barn. "I can't get any air," he said. Fortunately the does and chickens were secure and dry, thanks to the tarp with which we'd covered their leaky metal roof two years ago.

The Pygmies were already in the buck barn. The roof was not leaking, but the wind was blowing rain in through the windows and unchinked log siding. Because their shed is situated just off the crest of a small hill, water was streaming into it. Fortunately, the dirt floor at the back of the little barn was higher and drier, so we all huddled there until the storm passed. When the rain let up, Dan went to get towels to dry off the boys, as well as cardboard and dry straw for their bedding.

When we got back to the house, the power was out. We checked the rain gauge—1.25 inches in 15 minutes. The electricity came on about six hours later, but we lost internet service for days. On top of that, a nearby lightning strike had fried my computer. Usually when there is a storm, I turn off the computer and unplug it. This time, however, we were out the door before I could even think about it.

Four days later, internet service was restored to the area, but I still couldn't get online. While Dan chinked the log buck barn and dug a drainage ditch around it, I fiddled with a new Ethernet card. In the end, the billy boys had a snugger, drier buck barn and I had a new computer. We were all happy campers.

THE GOATS VERSUS THE BLUEBERRY BUSH

"Have you found your blueberry?"

Dan and I were out in the front pasture when a neighbor from across the street came over and asked us that question. We had been speculating about putting in a privacy hedge along the road frontage, when a voice called out,

"Have you found your blueberry?"

Not having a clue as to what he was talking about, we introduced ourselves. He responded with his name and then added, "I was just wondering if you found your blueberry."

"Our blueberry?"

"Yeah. There used to be a blueberry bush over there, by that sweet gum tree." He pointed toward a clump of trees near the center of the field where a lot of undergrowth needed to be cut back.

We told him no, we hadn't, and thanked him for letting us know. We chatted a little while longer and after we said goodbye, we went to hunt out that bush. I didn't figure any blueberry bush could have survived the years of neglect this place had seen, but when we went to look for it, there it was.

I immediately set out to liberate that bush. I cleared out around it to give it better air circulation and sun. It was loaded with blueberries just beginning to ripen. I got two quarts from my first picking and we had our first-ever homegrown blueberry pie.

The next year the bush did even better. Thanks to an end to our drought, the berries were bigger and sweeter, and there were more of them. Besides eating our fill on cereal, in pancakes, in muffins, for snacks, and for fresh pie and cobbler, I froze blueberries, dehydrated blueberries, and put them up as jam. What a blessing that blueberry bush was.

The following year we made a deliberate decision to sacrifice our blueberries that summer. Or at least to potentially sacrifice them. The reason? Our goats needed the pasture for grazing, so we decided to fence it in. I knew they would zero in on the blueberry bush, but their need was greater than ours, so it had to be done.

We did try to protect the bush by fencing the goats out. Dan put a ring of t-posts around it and attached welded wire fencing to it. Unfortunately, the welded wire offered the goats a perfect place on which to put their front hooves while reaching over to nibble on blueberry branches. Jasmine was quite the expert at that. The bigger problem wasn't just a matter of them eating the harvest, however. If goats take a liking to something they'll continue eating it down until they kill it. We definitely did not want that happening to our blueberry bush.

That was the summer we bought the two Pygmy bucks and moved the does to the back field once again. By that time, the protective blueberry fence was badly pushed in on all sides. Still, only the perimeter of the bush had been munched on, so not all of the blueberries had been lost thus far. We figured the Pygmies couldn't reach as high as the Nubians, so we took down the sagging circle of caved-in welded wire fence.

Things changed again as we began to work on an overall plan to divide our five acres for our future Kinders. The does had pretty much eaten what they wanted in the back field, and, according to them, were starving to death. We decided to do a switch-up, letting the Nubians back into the blueberry field, and giving the bucks a go at the back field. With the t-post and welded wire barrier now gone, it was an ongoing battle to keep the goats from eating down the blueberry bush. My goal: to save as many of the blueberries as possible. Their goal: to eat as much of the bush as possible, blueberries included.

I needed to do something and so we came up with the idea of a goat barricade made of pruned and dead branches. I used these to encircle my blueberry bush. This actually worked somewhat. Even so, it took no time for Jasmine to find the weakest spot in the barricade and make her way through, but at least it slowed her down long enough for me to come up with goat deterrent #3.

This was your basic, cheap, old-fashioned, no moving parts, lasts forever, fountain-type water sprinkler placed smack dab in the center of the bush. I turned it on every time I saw a goat anywhere near that bush, and at random times throughout the day. It worked like a charm.

When July 1st finally arrived, I anxiously went out to check the status of my blueberries. July is the month they're ripe and ready to harvest, and my mouth was watering for a fresh blueberry pie. When it came time for picking, however, the goats assumed that since the humans were allowed to touch the bush, so were they. I couldn't turn on the sprinkler while I was picking, but I was equipped with the next best thing: a squirt bottle.

One blast from that bottle changed their minds. I hooked it into a belt loop for easy access, leaving both hands free for picking. The harvest was so abundant that I was able to add canned blueberry pie filling to my pantry shelves.

When the branch barricade didn't work, I had to resort to more drastic measures. As long as the goats knew I had the squirt bottle, they pretended to mind their own business. Not so when my back was turned!

The best solution was the blueberry corral.

That winter we decided we needed something more permanent, so Dan made a rail fence from pine trees on the property. We call it the blueberry corral. It doesn't keep chickens and kids out, but they can't do much damage. The bigger goats can stand on the bottom rail and dream of eating blueberries, but the blueberries are safe.

I do give the goats the not-quite-ripe blueberries and I give the overripe ones to the chickens. The goats also get the branches when the bush is pruned. I dehydrate our abundance and add handfuls of dried blueberries to the goat feed as a nutritious treat during winter. So the happy ending to this tale is that everyone gets blueberries and there's enough to go around. This is homesteading at its finest.

The Game Changers

My first Kinder breeding attempts might have been a bust, but as long as I had at least two goats in my Kinder Starter Kit, I wasn't ready to give up. When breeding season arrived the following year, I continued to let Surprise visit Gruffy, but when she continued to go into heat it occurred to me that she might not settle (get pregnant) for the second year in a row. No kids meant no milk. I considered it a setback to not have kids, especially Kinder kids, but dairy had become an important part of our diet. I made yogurt, kefir, and cheese from our goat milk. I used it in cooking and fed it to the cats and dogs. We used the cream for butter and whipped cream on top of garden fresh strawberry shortcake. We needed milk.

A search on Craigslist rewarded me with a purebred Nigerian Dwarf doe in milk. Ziggy was a bargain, actually. She had taken a Junior Doe (maiden or not yet bred) championship and showed great promise as a show goat for her breeder. Unfortunately, once she kidded she did not have the perfect udder necessary to continue to win championships in the Senior Doe divisions. She was sold to me at a discount without her papers. That summer she provided us with delicious creamy milk.

I also needed do something about Surprise. There was little point in keeping goats if we didn't get kids and milk. If Gruffy couldn't do the job I needed a game changer. Perhaps it was time to re-evaluate my goals. The appeal of dual-purpose goats was that they seemed like another step toward self-sufficiency. Our acreage is too small to keep a lot of goats, and rather than keep enough goats from two different breeds to meet our needs, one all-purpose breed seemed the better way to go. As much as I wanted Kinders, I considered that there might be other possibilities for dual-purpose goats.

One possibility might be one of the heritage breeds. According to The Livestock Conservancy heritage breed goat chart, both the Arapawa and Spanish are sometimes used for dairy.[1] The Arapawa, a breed developed in New Zealand, descended from an old English dairy breed.[2] Either of these would be less productive than the commercially-developed breeds, but would be hardier and easier to keep. Unfortunately neither was available in my area. If I was going to import goats, it might as well be Kinders. I formulated a plan and bought two new goats. The first was an unregistered Nubian doe that I called Lily. The second was Elvis.

Elvis was a hunk of a buck, a purebred Kiko with a long silky coat and curly topknot that earned him his name by his breeders. The Kiko is classified as a meat breed and Dan thought them quite handsome. When I found a Kiko buck for sale, I bought him. I wondered if I couldn't develop my own dual-purpose breed. Perhaps I could breed Surprise to Elvis for a Nubian/Kiko cross, and Ziggy to Gruffy for a Nigerian/Pygmy cross. It certainly sounded doable. Surprise liked Elvis immediately and was fully cooperative.

Next, Ziggy went into heat. I knew she did, because she was standing at the fence flirting with Elvis. (If the girls aren't in heat they want nothing to do with the bucks.) I put the nix on that. She was a miniature breed and he was a fast-growing standard breed, so I didn't want to run the risk of her having kids too big to deliver. Instead, I put her with Gruffy. Was she ever mad! She took about three running charges and rammed him with everything she had. He was so ecstatic about being with a doe that he didn't care. Later I saw them lying together, chewing their cuds, as though they were the best of friends. Success? When it came time for milking, she couldn't get out of there fast enough.

About three weeks later, Ziggy went into heat again. Again, I dragged her in with Gruffy. Of course Elvis was watching and had much to say about it. To keep distractions to a minimum and everybody's mind focused on business, I separated these three by putting Gruffy and Ziggy in the back, in the fenced part of the woods,

All of my girls liked Elvis. He was a big, handsome Kiko buck whose topknot of curly hair had won him his name from his previous owners.

and Elvis in the front pasture. There were two fences and the buck pasture separating them. Satisfied with that arrangement, I went back into the house to paint the bathroom ceiling.

When I went out later to check on things, I saw Gruffy in the woods, but he wasn't pestering Ziggy as per usual. And he was staring at something. I wondered what. A few more steps and I could see what it was. It was Elvis in the woods with Ziggy! Elvis had easily cleared two 4-foot fences to get to her and was now busy keeping Gruffy away. I could only guess as to what else had been going on.

Well, I went to fetch Ziggy, who had had enough of the whole thing and was ready for her grain. I had no idea who had actually bred her, if indeed she had been bred. I would just have to wait and see.

In mid-March Surprise gave birth to a handsome little buckling I named Alphie. I had decided that I was going to call my Kiko/Nubians, "Kikobians" and wanted a name that would indicate "first." I considered Alpha, which morphed into Alphie, so Alphie it was. Early in May, Lily gave me twin doelings. Ziggy gave birth to quadruplets on Memorial Day, two boys and two girls. Sadly one of the doelings was stillborn. None of Ziggy's kids looked anything like Gruffy, but everything like Elvis. They eventually grew taller than their mother, which was telltale as to who their daddy was. So Elvis was the sire and thankfully, kidding had been incident-free. Maybe my Kikobians could be a Nubian/Kiko/Nigerian Dwarf mix.

I kept Ziggy's surviving doeling, but sold her boys and tried to figure out what to do with the rest of the kids. I decided to keep Alphie and give him Ziggy's girl the following breeding season. To broaden my Kiko genetic base, I bought another Kiko, a buckling we eventually named Hooper. We called him that because several times I found him in the neighbor's yard. Dan and I thought he was jumping the fence and got to saying, "Hoppity Hooper's out again." What we didn't know then was that he wasn't hopping the fence; rather, Elvis had torn a hole in it. He had hooked his horns in the welded wire, pushing and pulling until the welds popped and the hole was big enough for Hooper to get through.

When Surprise went into heat in October, she had a visit with Hooper. I put her and Hooper in one pasture, and the other two bucks in another. Next, Ziggy went into heat. She still wanted nothing to do with Gruffy so I put her in with Hooper. Several hours later I went out to make rounds. I first went to check on Ziggy and her date. As I stood there scanning the pasture Ziggy ran by. Then Hooper ran by. Then Alphie ran by. Lastly, here came Gruffy, running as fast as his stumpy little Pygmy legs would carry him.

One-day-old Alphie, my first Kikobian.

What happened?! Turned out that the gate was slightly open. Hmm. I was positive I'd latched it properly, but apparently I hadn't. Now I would have to wait to see if Ziggy went into heat again. If she didn't, then it would be another guessing game of "Who's The Daddy?"

Around the end of the month Lily went into heat. She and Gruffy were standing at the fence, and she didn't even want to come in for dinner. On a whim I thought, hey, maybe I can get some Kinder kids after all. I couldn't actually call them "Kinders," because the term has been trademarked by the Kinder Goat Breeders Association. Lily wasn't registered so I couldn't register her kids, but it's the breed qualities I was after. Hooper and Alphie were busy grazing in the other pasture, so I put Lily in with Gruffy. It was late, so I left them together overnight.

The next morning I stepped out onto the porch and looked for Lily and Gruffy. First Lily runs by. Then Alphie runs by. Then Hooper runs by. And lastly, here comes poor old Gruffy, huffing and puffing to catch up.

I let poor Lily back in with the does and went to inspect the latch on the gate. Not only was it open, but some of the bolts were loose. Somebody had figured out how to open the gate! And the evidence pointed to—Alphie! He was the one goat that was on the "wrong" side of the fence both times.

Neither Ziggy nor Lily wanted anything more to do with the boys after their ordeals. I could only wait to see if that lasted, or if they went into heat again. If they remained disinterested, then I would have no clue as to who had sired their kids. If their interest picked up again, then I might have another shot at my plans. Assuming I fixed the gate latch, of course.

The following spring we had a great crop of kids, ten in all: four bucklings and six girls. We were at an all-time high of nineteen goats!

One thing is certain and that is that nothing stays the same on a homestead. Things are in a constant state of change, and Dan and I have realized we need to adapt to them. Critters don't always cooperate with our goals and plans, because they have opinions and minds of their own. I wasn't sure how things would work out down the road, but one thing I did know—the adventure would continue.

It Was Time for Elvis to Go

One day, sometime between my first and second Kikobian breeding seasons, I headed out to the buck pasture to check water buckets. Our big Kiko buck, Elvis, was standing by the gate. This shouldn't have been a problem, but because of his wild streak, I always kept my eye on him and tried to keep my distance. As I slid through the gate, he came up and took a swing at my leg with his head. He did, in fact, catch my inner thigh with one of his horns. It wasn't what I would call a gouge, but it did break the skin like a scrape through my clothing. That, added to a string of other things, was the proverbial last straw.

Goats have a social hierarchy by which they live. If they don't see humans at the top, they will challenge the human's relative position in that hierarchy, and even try to dominate them. Elvis, having been raised with a herd of meat goats, was rarely handled when young. At best he was friendly sometimes, curious always, but never tame. We worked on that and I thought we'd made progress. He'd sometimes let us scratch and pet him, but never for long. He was always leery, and had been doing some mock challenging ever since we got him. He did some rearing up, pushed occasionally, but never charged us. Because of all that, I took care when I entered any of the buck areas. He took advantage, however, when I had to latch the gate.

Elvis had also taken to testing the fences. I'd watched him hook his horns in the welded wire and pull. He managed to pop welds and loosen the fence from t-posts in several places. One day I watched him ram the fence in a spot I'd recently fixed. If the place he worked on was a splice connecting the ends of two rolls of fencing, it would create openings through which the bucks could escape. Hooper's adventures into the next door neighbor's yard were one such escape. Another time I found Elvis and Alphie in our own back yard. For several days I was mystified as to how they were getting out, until I discovered their secret Elvis-made opening in the fence behind the bushes.

Elvis's destructive streak was not limited to fences. He trashed the hay feeder Dan made for the boys, frequently pooped in the mineral feeder (I know it was him because he was the only one tall enough to do it), plus continually knocked over the cinder block I kept in front of the mineral feeder for Gruffy to step up on. Elvis was constantly challenging Gruffy and pushing him around. As a Pygmy, Gruffy was only about half Elvis's size and would try to fight back until he tired of it. Then Elvis would resort to chasing him around. We had to tend to Gruffy's broken scurs on more than one occasion. (Scurs are the remnants of horns that grow after kids have been disbudded.) Oftentimes Elvis wouldn't let Gruffy into the buck barn, especially when the hay feeder had been filled with fresh hay. The same thing happened with Alphie, and then Hooper. Elvis had become a real nuisance.

If one wants to include eggs and dairy in one's diet, then certain decisions must be made in regard to breeding and the resulting offspring. Chickens can lay eggs without a rooster, but for chicks the eggs must be fertilized. For milk, goats must kid and cows must calve, which means a male must be available. If one wishes to keep and breed female offspring, then a sire other than their father is desirable. Those with large acreage may not have to address these concerns, but for those of us with small holdings, the question of how many animals can be kept is vital. Unfortunately, it is not simple to answer.

For Dan and me, the primary goal for our homestead is sustainability. We want our homestead to achieve and maintain a balance. We believe it is an essential part of stewardship. The number of animals we can keep begins with the soil. Healthy soil means healthy plants, which means healthy animals. It is a balance from which we all benefit. We started with very poor soil and have only been able to rebuild it slowly. Every year we choose one pasture area to test and amend. We address not only organic matter, pH, calcium, magnesium,

Hooper, Elvis, and Gruffy. Elvis was the largest and dominant buck.

nitrogen, phosphorous, and potassium, but the micronutrients as well: boron, copper, iron, manganese, molybdenum, and zinc. The test for these is more comprehensive than a standard soil test from the state cooperative extension service and more expensive. Natural soil amendments are more costly than chemical fertilizers, but we consider it an investment in our homestead and in our future.

In spring, things begin to grow and young are born. They thrive during summer when we are at peak forage. We keep an eye on that forage to make sure it is not overgrazed and eaten down. Toward that end we practice pasture rotation. Winter is the leanest time, the time by which we need to have trimmed down our numbers according to how much our winter pasture can support and how much feed we have stored. How to do that trimming is something every keeper of animals must address. From the beginning we decided we would raise goats for milk, manure, young, to trade or sell, and meat. That means the extra animals are not unwelcome.

Females are usually easy to trade or sell, males are not. I have found purebred breeds, registered or not, easier to sell than crossbreeds. First I would try to sell Elvis, but if I couldn't we had only one choice left.

The choice to raise and eat one's own meat is not an easy one and, in fact, can be controversial. For those new to raising animals, it is difficult when an emotional bond has been formed, because it is nearly impossible to imagine eating something that has been so cute, friendly, and entertaining. For others, it is a moral issue.

There are several faces to the morality of eating meat. One is that it involves taking the life of the animal. The objection to this may come from religious convictions, or as an animal rights issue. The second face revolves around the inhumane treatment of industrialized agriculture toward production animals. The abuses are sickening. For example, CAFOs (Concentrated Animal Feeding Operations) which crowd livestock into confined, vegetation-free areas for weeks at a time with no waste removal; commercial egg factories which increase production through forced molting by starving close-caged chickens; or the disposal of unwanted young in egg and dairy production by discarding them alive in Dumpsters. Is it any wonder so many are vehement in their choice not to eat animals or use animal products?

This begs the question, is veganism the only alternative to such abuses? For those who rely on industrialized agriculture to produce their food it probably is, because industrialization has been the primary cause of mistreatment and abuse of livestock animals. Animals are treated as production machines with profit as the only objective. But what about those of us who work on a much smaller scale and choose to raise a few animals for a multiple of reasons? Or smaller family farms which treat their animals humanely? Is it right or fair to lump all into the same category? Is it logical to assume that people who eat meat hate animals? Obviously, I will insist that it is not, while others will stick to their position that it is.

It is interesting that former vegan Lierre Keith points out that most of the facts in support of vegetarianism are based on the industrial farming model. For example, the argument that the twelve to sixteen pounds of grain it takes to make a pound of beef could be better utilized in feeding people. For CAFO beef, this is true.[1] The development of sustainable agriculture, however, is changing the point of reference. Keith points out that Joel Salatin uses no grain to produce beef because his cattle are pasture-raised. Salatin, in turn, points out how pasture-raising livestock is a much more efficient use of the world's resources, plus better for the environment. In *You Can Farm*, he states that approximately 70% of the grain grown in the US is fed to livestock.[2] Think about it. How would switching to no-till pasturing change things for the animals and the environment? The point I am trying to make is that there are alternatives to the abuse, and that an increasing number of folk are utilizing them.

Is it possible to homestead or farm without animals? Is it possible to support oneself or a community from the land without them? I think that's another valid question. Scott and Helen Nearing were certainly successful at it. They established and maintained self-sufficient homestead living for nearly 60 years without animals. In a very low-tech, hands-on approach to their food, they managed to feed themselves year-round on a strictly vegan diet.[3]

Not everyone has the time, land, health, and inclination to grow all of their own food, however. Is it possible to grow enough food without animals to feed a community? Or stock a grocery store? An interesting article at the Windward Education and Research Center website discusses just that.[4] Windward community member Andrew Schreiber, a former vegetarian, makes a rather thorough examination of the role of animals in creating and sustaining human communities. He concludes that veganic permaculture might be possible in tropical, subtropical, and coastal maritime climates. There, the diversity and abundance of plant life is adequate to meet human nutritional needs, plus build and maintain high soil fertility. Without that, however, the only non-animal alternative is the use of synthetic soil nutrients derived from fossil fuels, because natural sources are primarily animal products: manure, blood meal, bone meal, and fish meal. Schreiber further points out that the social, political, economic, and ecological impacts of utilizing petroleum in this manner; neither benefit animals nor eliminate their death and suffering.

How so? Because large-scale production of field crops displaces entire ecosystems by requiring large amounts of cultivated ground. Without even native animals to maintain soil fertility, the only option is synthetics. In other words, trying to eliminate critters from food production only makes the problems worse.

I believe that how we now view animals is a product of our culture. That culture is no longer an agrarian one, it is no longer based on meeting our needs from the land, but on meeting them through industry and advanced technology. This has created a mental, physical, and emotional disconnect between humans and the natural world. We are no longer invested in how we feed and support ourselves; instead we rely on technology, industry, and the big businesses they represent to do that for us. That leaves us free to explore other pursuits, create leisure and entertainment, and be all we can be. And yet, there is an underlying dissatisfaction with modern life as we know it. We see that there are problems and we want to fix them. We try to do this with politics, causes, and contributions to causes. Yet because of that disconnect, the problems are never solved. In fact, other problems are created. We can try to save animals by refusing to eat meat and use animal products, but the only substitute for them is found in fossil fuels. Whether in agricultural chemicals to grow our vegetarian diet, or the faux fur and looks-like-real leather for our garments and footwear, we must rely on the unsustainable petrochemical industry. We find ourselves in a landslide of environmental and economic disaster, desperately trying to rearrange rocks and rubble while we plunge to our doom.

Life requires death. We may deem animals more worthy of life than plants, but it is death which sustains life and that is something we cannot change. All natural ecosystems include animals. Human activity modeled on natural systems will include animals as well. To remove them would be to create an artificial ecosystem. The challenge of being part of a natural ecosystem is to know our place as stewards and servants of our land and all it sustains.

So what did we do about Elvis? First, I listed him on Craigslist. I wasn't especially hopeful about this option, because Craigslist is always overloaded with ads for bucks. Bucks, even the little guys, are hard to sell because there are so many of them. After about a month, I had one inquiry from someone who wanted a buck to breed their does, but they wanted one which would be safe with children. I had to tell him that wasn't Elvis. (No buck is actually. Even Gruffy, who had a gentle and affectionate personality most of the year, was a different animal in rut. I gave him a wide berth then and never turned my back on him.)

Our options for Elvis had been narrowed for us. He was too destructive to remain on our homestead, and no one wanted to buy him. Our decision was no longer what to do with him, but how. In some areas it is possible to hire someone to do the killing, dressing, even hauling off to the butcher for cutting and packaging the meat. In our area no such service is offered, which meant we had to decide whether to take him to a meat processor or do the job ourselves. This is the question we discussed. When Jasmine's broken shoulder never healed properly, we opted for the processor. There were emotional reasons for this, but also, we weren't ready for that yet. However, it had always been something we'd planned to do eventually. It's tidier to have someone else do it, both physically and emotionally. On the other hand, there's a lot of waste. Dan hates waste. The bottom line, however, is if our goal is self-reliance and we eat meat, then we should take responsibility for it.

In the end we got the proper equipment and did it ourselves. I hate to say it, but neither the bucks nor the humans missed Elvis after he was gone. I'm sure the billy boys were happy not to be picked on any more. Dan and I simply appreciated not only the meat, but the peace and quiet as well.

HOOPER'S CLOSE CALL

"Rut" is the male goat's counterpart to a female's heat. A buck in rut is a rather disgusting creature, at least to civilized sensibilities. If you are reading this to your children, unless you're raising them farm-fashion, you might want to:

    a. cover their ears,

    b. make up your own words for this part, or

    c. skip over the next paragraph altogether.

Like any other male wishing to make himself appealing to ladies of his kind, a buck in rut will do what it takes. He will urinate all over himself, especially his face and beard, even into his own mouth. This is allowed to dry and "age" which only heightens his perfume. He may think you find this desirable and spray you as well. He also loves showing off his erection, to let everyone know he's available. He will mouth his male member too, as though very proud of it.

With his hormones raging, an ordinarily tame and friendly buck is a changed personality. I would never, ever, turn my back on a buck in

Alphie was a friendly guy, but he would use his horns to his advantage.

rut. I wouldn't turn my back on a buck anyway, without keeping one eye on him, but a buck in rut is about one thing and one thing only: mating. If a human is seen as in the way or even a rival, watch out.

When not trying to attract the ladies' attention, bucks in rut will spend their time sparring with one another. It can mean bloody horns and scurs or other injuries. Or, as in Hooper's case, it can mean other potential hazards.

Hooper was the young Kiko buckling we bought early one summer to breed our Nubian and Nigerian does in the fall when they came into season. Past experience had taught me not to wait until the last minute to find a buck, so I would start looking late in the spring or early summer for the next season's buck. Hooper was Elvis's replacement. At the time of this tale we had Gruffy, our mature Pygmy buck, and Surprise's buckling by Elvis, Alphie. Alphie and Hooper were about the same age.

One night Dan was out standing on the back porch after dark. The neighbor's dogs were barking like crazy, but he told me he heard something that really got his attention—not a dog, but a goat. It was the sound of a goat in distress. He took the flashlight and headed toward the buck shelter.

When he got there he found a mess. Hooper and Alphie had gotten into a tussle. Both have horns and both have collars. Somehow Alphie had gotten a horn hooked under Hooper's collar. In their struggling they had gotten completely tangled up; Hooper was strangling and Alphie was panicking. Dan said they were so twisted up that Hooper's collar was too tight to get off. He flipped Alphie head-over-heels to loosen the collar. He could not get Alphie free but was finally able to release the catch on Hooper's collar to free them both.

He said Hooper stood there dazed, pupils fixed, glassy-eyed. Dan kept calling his name, trying to get him to respond, but Hooper went down and stopped breathing. Dan started furiously pumping his ribs and blowing into Hooper's nostrils. Hooper finally came to.

You goat folk are probably thinking of the same two things I am—horns and collars. Horns are a topic of controversy amongst goat owners, and there are many "war" stories out there about them. I have a few of my own.

There are two terms referring to horn removal which are often interchanged but actually mean different things: disbudding and dehorning. Disbudding refers to addressing the issue while the kids are still very young, before the horns have started to grow. The horn bud is burnt with either a disbudding iron (unfortunately often called a dehorning iron, which confuses the issue) or with a caustic paste. Very few vets and goat breeders recommend the paste. The disbudding iron must be used carefully, to only destroy the horn bud and not do further damage to the kid.

Dehorning is the removal of horns once they are growing. It is a surgical procedure that many, many vets refuse to do. There is an ample blood supply in goats' horns (they are, for that reason, warm to the touch). This means that dehorning is a very bloody procedure for which bleeding must be controlled. As you can imagine, there are horror stories out there about folks who thought they could simply saw them off.

Initially I was in favor of disbudding, but Dan was not. He is very much against any number of livestock management techniques that he doesn't consider natural. Besides that, he likes the look of horns. We spent several weeks discussing what we would do when our first goat kids were born.

While dairy breeds are traditionally disbudded as kids, meat and fiber breeds are not. Dairy goats headed for the show ring are disqualified if they have horns. Personally, I have had too many bad experiences with horns to like them. Dan, on the other hand, thinks they are natural and beautiful, plus a goat with horns can defend itself from predators. A lot of folks feel that way and would never disbud their goats. Others won't leave them with horns because, for example, that same predator can kill that goat if its horns are stuck in a fence.

I've purchased goats with horns and goats without, both disbudded and polled (naturally hornless). One of our first goats, Abigail, used her horns to bully Surprise when she was the newcomer. One time she drew blood. Plus, Abigail would get them caught in the hay feeder and get stuck (which proved useful for giving her vaccinations or trimming her hooves).

*Elvis got his head stuck in cattle panel fencing more than once,
but because of his horns, he couldn't get out. These were the
only times I could trim his hooves and give him his vaccinations.*

Then there was Petey, my "Plan E" buckling. He would use his
horns to push everybody around, including us, and including my llama
Charlie. A couple of times I found clumps of Charlie's fleece caught in
Petey's horns. If he wanted something in your hand—something to
eat, for example—he would try to grab it with his horns. Needless to
say, that guaranteed that Petey would either get a new home once he
finished his job or would become sausage.

Probably Elvis and his offspring Alphie were the worst when it
came to using their horns. They would often swing their heads at me
and occasionally caught me with their horns. It was never a deliberate
attempt to harm me; it was just head-swinging to either get something
they wanted or to show uncooperativeness with whatever I was trying
to do. The scrapes and bruises I sustained did nothing to win me over
to the horned goat camp.

When we finally decided we would indeed disbud our first kids,
Dan looked online for more information. Unfortunately, he found a
couple of how-to gone wrong videos on YouTube. One such video
showed someone burning a kid's skull all the way through to the
sinuses. Well, that made Dan furious. More research indicated that we

had waited too long anyway. The kids were about three weeks old then and they really should have been disbudded within a week to ten days, or as soon as the buds could be detected.

Hooper, being a meat breed, wouldn't have been disbudded anyway. This brings us to the other controversial topic: collars.

Most goat owners agree collars are a necessary evil. Many folks like plastic chain-look collars that come in different colors. These are said to be fairly sturdy but will break apart if necessary. I doubt they could be used to catch and hold a goat that is adamant about not wanting its hooves trimmed. Or to drag away the goat who has managed to dart through the gate at feeding time and jam her head into the open bag of sunflower seeds.

It was fortunate that it was Dan who found Hooper and Alphie in their tangle. It was fortunate that he happened to go outside at that moment, and that he was even home at the time. It's unlikely I would have had the strength and weight to come between the two struggling bucks to free Hooper. But what could have been, wasn't, and I'm very thankful for that.

Hooper remained collarless until I sold him. He hadn't been handled much as a kid so he was never cooperative when it came to being caught, and without his collar, he was harder to catch than ever. For hoof trimming, Dan learned how to use a lasso to catch him. Not elegantly perhaps, but it did the trick. To his credit, Hooper always settled down once caught, making him easy to handle if needed.

The experience with Alphie and Hooper did not change Dan's mind about horns and disbudding, and until recently, our homestead-born kids have had horns. I eventually bought a disbudding iron and learned how to use it. Our goats still have collars, but thankfully nothing like this incident has happened since.

GOAT BUSTER

One thing that can be said about goats is that they are predictable. They will predictably get into whatever you don't want them to; the incident with the blueberry bush, for example. Fencing now keeps the big goats out, but the little ones are still able to crawl under it! Another problem area has been the chicken yard.

Our chicken yard has a gate which allows the chickens out into the pasture areas to free range. At night I close the gate to keep predators out. The problem is that the chicken yard is irresistible to goat kids. Every kid we've ever had on the homestead ends up in that yard! And not only kids, but miniature goats too.

Edy and Nessie were two of my Nigerian Dwarf does. They weren't sisters but had been raised together. They were the only goats I've ever had that I could feed from the same pan without one chasing off the other. Oh, there were some mock standoffs, but neither wanted to risk the other getting more than her fair share, so they shared the pan. They did everything together and where one went, the other followed, including accepting the intriguing invitation of that open chicken gate.

This might not have been too bad, but goats that made it through the gate invariably found their way into the chicken coop as well.

Our chicken yard gate proved to be an irresistible temptation for small goats.

Once they realized there was a chicken feeder full of feed in there, it would become impossible to keep them out.

I had tried everything I could think of to keep them out. I narrowed the opening with boards but if a chicken can get through, so can a kid. I once constructed an elaborate tunnel for the chickens to pass through to get to the gate. Instead of deterring the kids, it gave them something fun to do!

The only thing for it is a steady blast from the garden hose. It works wonders at chasing them out. Problem is, I don't always have the hose available and at the ready. Then one day I found just what I needed on the clearance shelves at Walmart: a pump action, 30-ounce capacity water gun with blasting power up to 40 feet. And for only $3! How could I go wrong? The goats suddenly realized that the chicken coop wasn't nearly as interesting as they had originally thought. A win-win for both me and the chickens.

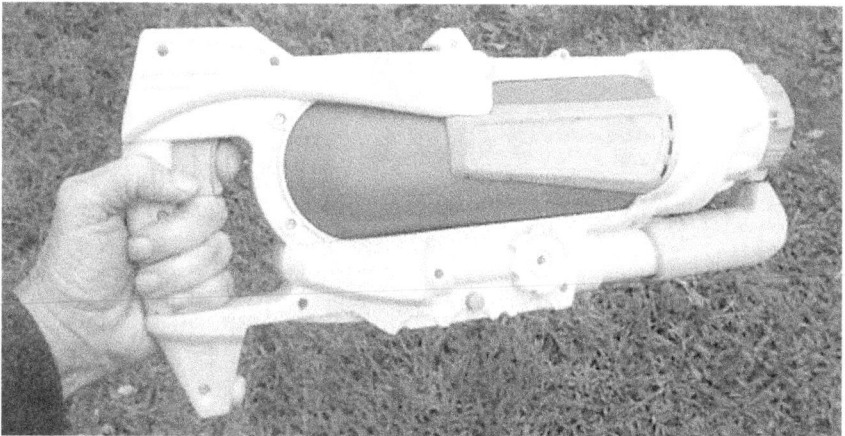

My goat buster. Water seems to be the best goat deterrent there is.

The best way to keep kids out of the chicken yard is with a section of cattle panel. The chickens can go through the panel openings, but kids cannot.

GOAT MIDWIFERY:
LEARNING THE HARD WAY

I think kidding is one of the most exciting times of year on the homestead. New baby animals are always a joy, and baby goats are exceedingly entertaining. Anticipation runs high, as does lack of sleep from frequently checking on whichever doe has the next due date. Most of the time things go well and very little assistance is required, but sometimes there are problems. Such had been the case with Jasmine (see page 117). Her stillborn kid was the only problem I encountered my first year, but it left me a little nervous about our second kidding season. That was the year I tried to breed my Nubians to my Pygmy but ended up kidless. It was the year we lost Jasmine, and it was the year I gave up on Kinder goats.

   With a new goal and new bucks, I was hopeful for a better outcome. Surprise was the first to kid that following spring. I kept a close eye on her, especially once the due date circled on my calendar came and went. When her udder started to fill I knew we were close.

The next day she stayed standing in her stall and wouldn't come out, not even for a brief nibble in the new pasture. At 9:08 p.m. I went out to check on her and discovered that she had started pushing. Shortly, the birth sac appeared. I looked for the tips of the kid's two front hooves but saw none. After half an hour, there were still none, and no progress. This was worrisome and I dreaded another difficult birth. They're only supposed to happen in some 90-odd% of kiddings, and this being my fourth, meant my average was becoming 50%!

I went to my birthing kit and got out a glove and lubricant to go in to "see" what was happening. Surprise was not at all cooperative and would not stand still. At last I was able to feel two legs and teeth. That's a normal headfirst presentation, but what was the problem? I realized that what I was feeling was not feet but knees. The kid's legs were bent at the knee, instead of heading straight out the birth canal like they were supposed to. There wasn't enough room to get my hand farther in to feel anything else. Under normal circumstances, a doe should deliver within 30 minutes, but it was going on an hour of hard labor by now and that was too long. I prayed and managed to guide the front feet out. With the next few contractions I pulled. Surprise screamed and was finally able to get the head out. A wet, limp baby boy soon followed.

He wasn't moving, he wasn't breathing. I grabbed my towels to wipe away the birth sac from his face and nose and began rubbing vigorously. Surprise licked him like crazy. Finally he moved and whimpered. More rubbing and licking and he began to complain. I was so relieved.

This was the little guy we named Alphie. My kitchen scale weighed him in at 10.5 pounds. Besides being large, he already had tips of horns showing! Of course, he was born at about 158 days' gestation; average for a goat being 145 to 155.

I made sure there were no other kids coming and that the placenta was out. There was no need to dispose of it because Surprise ate it (not uncommon with goats). I also made sure the little guy was dry, had a tummy full of colostrum, had his selenium/vitamin E paste, and could wobble about and find a teat all by himself. Because our temperature was supposed to dip back down below freezing, I hung the heat lamp in the stall and finally got to bed around 2 a.m. I went back out at 5 to check. Mother and baby were doing fine.

Likely most goat breeders would tell you that the heat lamp wasn't necessary; that as long as he was dry and nursing he would be okay. I worried though, because I often read about other bloggers losing kids from hypothermia. I'd had enough problems with Surprise and I didn't want more.

All those sad stories of death from hypothermia are the reason I tend to breed my does later in the season. I aim for kidding in late March or preferably April, when the worst of our cold weather is past. So why do folks breed earlier, so that the kids will be born during freezing winter months? They do it when they've had trouble with persistent coccidiosis, another kid killer.

Coccidia are protozoa which occur naturally in small numbers without causing a problem. Most goats, in fact, develop an immunity to them as they mature. Kids acquire coccidia by mouthing things, so that the organisms invade the digestive system and rapidly reproduce to cause severe scours (diarrhea), irreversible intestinal damage, and death. Emergency treatment is sulfonamides given orally, and it requires immediate action to save the kid. Early kidding helps prevent this because the temperatures are too cold for coccidia to be active.

Fortunately I had neither of these problems with Surprise's Alphie. What he did have, however, was contracted tendons. This is where a newborn appears to have leg problems; usually the leg won't straighten and the baby can't get around without limping or hobbling. Many folks will attempt to splint the leg to correct it, but my vet said it would straighten out on its own, which I have found to be true. He said it occurs when a kid lies in one position in the womb for a long time. In Alphie's case, his front knees had remained bent before he was born. Even though I was able to straighten them out for his birth, they remained in a partially bent position for the first several weeks of his life. Contracted tendons rarely stop a bouncy kid, however, and the limbs straighten from exercise as the kids begin to grow.

Two months after Alphie was born, Ziggy gave birth to triplets. She had never had problems during kidding, except for a stillborn doeling the year she had quadruplets. This time, however, we had a new problem: retained placenta.

Usually a doe expels the placenta within 24 hours. Sometimes she will eat it, or part of it; other times it must be removed from the stall and buried. Ziggy only partially delivered the placenta, so that it was hanging out and dragging on the ground. I was worried about uterine infection so I trimmed it with a pair of sterilized scissors and took her to the vet. There are many goat problems I felt comfortable handling myself, but this was not one of them. The vet was able to remove it and gave her antibiotics to prevent infection. She also gave me some very good advice: rather than cut it, knot it. This will get it off the ground and the weight will help pull it out.

I was sent home with oxytocin to stimulate the contractions needed to expel anything that remained. We also got a week's worth of Naxcel injections, an antibiotic for which there is no milk withdrawal.

Multiples are common for Nigerian Dwarfs. Ziggy was an excellent mother.

Most medications (such as antibiotics, chemical wormers, etc.) have a withdrawal period. This is the time period in which traces of the medicine can be found in the doe's milk or muscle. The milk or meat is withdrawn from use and discarded until that time has passed (although I've saved such milk for soap making).

My other concern with Ziggy was whether or not the triplets were getting enough to eat. After they're born, I always make sure each kid finds a teat and gets a tummyful of colostrum. However, Ziggy's udder never filled well. It was also very saggy and low to the ground, adding the additional challenge for the kids of finding it! Fortunately, I keep colostrum in the freezer and was able to offer some to each kid.

Colostrum is one of the three dairy products I freeze (the others are cream and grated goat milk mozzarella). Colostrum is that nutrient-rich first milk, or perhaps pre-milk is a better description. It provides the kids with maternal antibodies, helping to give them a strong and healthy start in life. It's possible to buy generic, powdered colostrum at the feed store, but after looking at the ingredients, I was convinced to save and freeze real colostrum from my does. My Nubians especially produced more milk than their newborns needed, so I would milk out a little, pour it into muffin tins, and freeze to store for future emergencies. It can be removed from the tins, kept in freezer bags until needed, and then heated and bottle fed to any kid that needs it.

Eventually Ziggy's little doeling caught on to nursing, but the boys, not as well. Baby goats do not always know where to find milk. They have the instinct to suck and they know mom is the source, but finding that teat with its reward of warm tasty milk is sometimes a hit-or-miss learning process. I'd frequently try to guide each kid to a teat, but for these boys, there didn't seem to be enough milk to satisfy them.

I started supplementing their feedings with a bottle. I started with 100% colostrum and gradually mixed it with milk from another doe. The transition from colostrum to milk varies, most say in about a week.

I named the doeling Zoey, and she seemed to get enough of Ziggy's milk. She refused the bottle, but the boys latched on to it with great vigor. Eventually the bigger of the boys caught on to nursing. He was a stocky little guy that I named Buster Brown. The littlest was always pushed out of the way and likely wouldn't have made it if I hadn't intervened. Ziggy would look at me as if to say, "I'll take care of him but you have to feed him." He was the last born that season, so I named him Zed. He would come running for his bottle when I called, "Where's my baby?" Bottle feeding is more work, of course, but it's just one of those things which must be done on occasion. Fortunately, I kept the recommended Pritchard nipples on hand, so I was ready to bottle feed in an emergency. I have to admit that bottle feeding a baby goat is a pretty sweet experience.

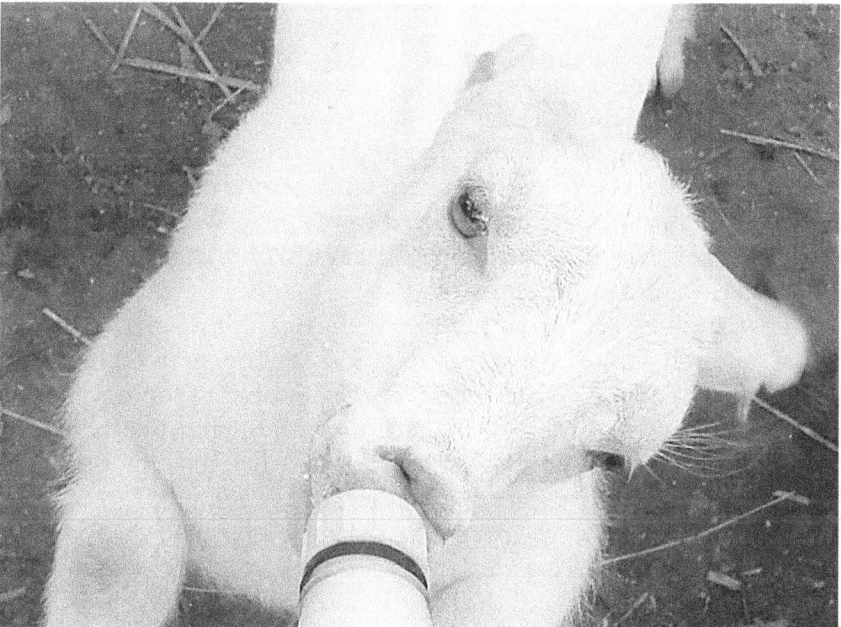

This is Zed, my first bottle baby. As the triplets got older I began offering the bottle to his brother Buster Brown as well. Ziggy was not a big milk producer, and her growing boys needed more than she could produce.

When my third kidding season arrived the following spring, I felt a little more confident for any problems that might arise. Even so, it started out with something totally unexpected.

One afternoon Surprise was standing near the barn with the spaced-out look she got when she was in labor. She was just a couple of days past her official due date so I figured this was it. I had to do a quick kid shuffle, putting recently-delivered Ziggy and her new set of triplets in the larger far stall, and Surprise in the nearer kidding stall. I made sure everyone had plenty of clean bedding and then hung around to keep an eye on things.

Several hours later Surprise lay down. There was no discharge and no pushing, so we waited. And waited. The longer we waited, the more disoriented Surprise became. She became unresponsive to my voice and could not stand up. Something was clearly wrong and I had the horrible feeling that Surprise was dying. Since my vet didn't make house calls, and I couldn't get her into the back of my Jeep by myself to take her to him, I knew I was on my own. I did a quick internet search of "pregnant doe lethargic." There was a lot of information to sift through, too much, really, for my frantic state of mind.

Surprise's symptoms fit those of pregnancy toxemia. The best emergency information I could find (i.e. specific, bulleted instructions that I could follow easily without digging through volumes of other information) was at Molly's Fias Co Farm website.[1] I followed her treatment recommendations, starting with 20 cc of molasses and corn syrup mix every two hours by mouth. Also I gave Surprise subcutaneous injections of calcium gluconate (40 cc total over 4 sites with 10 cc each location). I tried the recommended Nutri-Drench too, but she really fought me on that one. The next time I went to check on her she was alert and turned her head toward me. Within a couple of hours she was back on her feet and I gave her a vitamin B-complex injection to stimulate her appetite.

It was a relief she was okay but there was still no sign of labor. More research indicated that unborn kids rarely survive pregnancy toxemia, so I had to prepare myself for the devastating consequences. I checked on her every several hours during the night. She remained standing and shifting her weight like she does when in labor, but there was nothing more to indicate that kids were imminent.

A little after 6 a.m. I got up, started the coffee, and headed back out to the goat shed. There, on the ground, was a live black and white baby! I ran to get Dan and shortly after we got back a second live kid was born. Both doelings! A miracle and a blessing all in one.

I continued with the calcium injections for several more days. Once my mind was calm again, I sat down to do some serious research on

Surprise and her miracle twins. Finding a veterinarian who is knowledgeable about goats and willing to make farm calls is not easy.

exactly what had happened and why. I concluded that although my initial diagnosis was wrong, the treatment was nonetheless correct.

There are two late pregnancy problems for goats (and sheep) which are life threatening: pregnancy toxemia (also called ketosis or twin lamb disease) and hypocalcemia (milk fever). If not addressed immediately the doe or ewe will die. Neither are diseases, but rather, metabolic conditions which are primarily feed related. The cause and treatment are different, but since hypocalcemia can lead to ketosis, treating for both is a good idea anyway. Ultimately, the response to treatment determines diagnosis.

The symptoms are nearly identical:

    loss of appetite

    lethargy

    weakness

    disorientation

    goes down and can't get up

Pregnancy toxemia occurs when the body's demand for energy (carbohydrates in the form of glucose) exceeds what the diet provides. It usually happens in late pregnancy, when the kids are rapidly growing. If the dam isn't consuming enough carbohydrates to meet the need, her body begins to metabolize fat for energy. Ketones are the byproduct of fat metabolism. As they accumulate, the system becomes increasingly acidic to the point where it is fatal. This can happen to people too, when they do not have sufficient insulin to metabolize their intake of carbohydrates. This is referred to as diabetic ketosis.

For goats, treatment requires immediately supplying energy until the doe begins eating on her own. This can be a commercial product such as Nutri-Drench or Goatdrench, or propylene glycol (available at feed stores for this purpose).

I admit to having negative feelings about using propylene glycol. It's the "antifreeze" used in most commercial ice creams. In an emergency, however, it's better than losing the dying doe and her kids, because it is easily assimilated by the body, immediately providing much needed energy. The primary objective here is life rather than death. Propylene glycol is the primary ingredient of Nutri-Drench and Goatdrench, which also contain molasses; calcium; vitamins A, D, and E; and selenium. Alternatively, a homemade drench can be made from one part molasses to two parts corn syrup.

Also important are B vitamin injections to stimulate appetite; probiotics, yogurt, or kefir to reestablish digestive flora in the rumen; and water, force-fed if necessary. The doe needs water to begin flushing the ketones out of her system.

Pregnancy toxemia is what I initially thought was Surprise's problem. But I also noted that the website said that the milk fever treatment also helped.[2] Because of that, I gave the calcium gluconate injections. In an attempt to find anything for her to eat, I offered dried comfrey leaves. Comfrey is rich in calcium and because it was the only thing she was interested in (she devoured it, in fact) this was the clue that helped me later diagnose Surprise's actual problem.

Hypocalcemia occurs when the doe's diet contains an improper calcium-phosphorous ratio. Grain is usually the culprit here, because it is high in phosphorous but low in calcium. The pregnant doe needs at least twice the calcium in her diet, i.e. two parts calcium to one part phosphorous, especially during the end of her pregnancy, when the kids are needing calcium for bone development, or immediately after kidding when her body begins to produce milk. If she isn't supplied calcium through her diet, her body will begin to deplete her own resources. Most folks know calcium is necessary for healthy bones and teeth, but it is also essential for muscles and nerves to function properly. Skeletal, heart, digestive, and uterine muscles all rely on calcium. With a calcium shortage, muscles become weak so that the doe becomes weak: she can no longer stand, digestion slows, uterine contractions will weaken, and eventually her heart will give out if the problem isn't corrected immediately.

The treatment requires 40 to 50 cc calcium gluconate injections subcutaneously (just under the skin). That's a lot, so this must be divided into four or five 10 cc doses and injected slowly in four different spots. This is repeated hourly for the next two hours, and then several times a day for several more days. Because the doe has stopped eating, ketosis will be a secondary problem which must be addressed as well.

So why did this happen to Surprise and what could I have done to prevent it? Problems and prevention are all about diet. Some commercial feeds have the correct calcium-phosphorous ratio. I had avoided these because they rely heavily on corn and soy, two crops which are commonly genetically modified in the U.S. I fed one scoop of whole wheat and oats, with two scoops of alfalfa pellets for the calcium. When I learned that my alfalfa source had started using genetically modified alfalfa, I became alarmed and stopped buying it. This was my near fatal mistake, because I did not have a sufficient calcium replacement.

Besides alfalfa, possible other sources of calcium include dolomite powder, as recommended by Pat Coleby.[3] Also comfrey is rich in calcium. Another possibility is Chaffhaye, for those who can get it locally (or can afford to have it shipped). Chaffhaye is a packaged, fermented alfalfa feed product, known as "pasture in a bag."[4] Alternatively, don't feed grain. Grasses and hay usually have the correct balance of calcium and phosphorous. As long as the goats' mineral mix is balanced, additional calcium isn't necessary.[5]

The incident with Surprise was more than enough excitement for one kidding season, and I was fortunate that things worked out well. Losing animals is always difficult, making victories all the sweeter. Needless to say, calcium gluconate and Nutri-Drench are now standard items in my birthing kit. What is interesting is that my other three pregnant does were on the same diet but did not have the same problems. We did, however, have another problem.

Lily, a pretty moonspotted Nubian, was the next doe to kid that spring. This was Lily's second kidding. A doeling was born first. I was pretty sure there would be another, and sure enough, a second birth sac bubble soon appeared. I was glad to see two tiny hooves, which indicate a normal presentation. Lily pushed while I watched and waited. Going seemed slow and the feet would appear, disappear, and reappear. I was getting a little anxious, when, with the next contraction I saw four tiny goat feet! Yikes! What was going on? I slipped my hand in along beside the feet and realized that two kids were trying to be born at the same time. With her next contraction I grabbed onto the pair sticking farthest out and pulled. Lily screamed and gave birth to a huge baby boy. Another little doeling soon followed and all was well. What a relief.

Last to kid that year was Zoey, a first freshener (a term referring to a doe kidding for the first time. Freshening refers to coming into milk after kidding). Zoey surprised me, too. I was doing routine checks on her, knowing she would kid soon. When I gathered up the goats for evening feeding and milking I noticed Zoey wasn't there. I scanned the pasture and saw her in the far corner. Was everything all right?

If she'd isolated herself, it could only mean one thing. As I got closer, I could make out something small and brown standing next to her. I picked up my pace. Then I saw something small and white on the ground behind her. I started running. Twins! Not only that, but they were both completely dry and she'd already delivered the placenta. I was shocked I'd missed it but relieved everything was all right.

It was hard getting her to leave that spot, even though I was carrying her twins and sticking them right under her nose. We finally got to a clean, comfy, safe stall and I got the new little family settled in. She had a boy and a girl, and we wrapped up kidding season for the year with ten healthy kids: four bucklings and six doelings.

Some folks have the benefit of a nearby vet who is knowledgeable about goats. Many of us do not and, more often than not, find ourselves trying to manage difficult situations on our own. As with so many things, I've learned more by doing than by reading. Not that researching, reading, and study aren't important, but I find it impossible to retain all the bits and pieces of information I need in an emergency. What I do find, is that something in my subconscious memory will nudge me in the right direction. I've had my losses, sometimes because of my lack of knowledge and experience, and sometimes in spite of doing all the right things. There are no guarantees about life and death. Dan and I have learned to accept them, learned from things that don't turn out as we hope, and rejoice in the things that do.

SOLVED: MYSTERY OF THE DYING KIDS

I hate learning things the hard way, but for several years I had been bewildered by the mystery deaths of three young bucklings. I had these boys in different years and they were different breeds. Two were purchased from breeders in different states, and the third was born on the homestead. No other goat had been or became sick. None of them had diarrhea, which is a symptom of coccidiosis—a dreaded disease which often afflicts young goats. What they all had in common was that they died shortly after they'd been weaned at two months of age.

The symptoms were the same: weakness, lethargy, and then death within just a couple of days. The first one I tried to treat myself. I searched the internet and found the symptoms matched a disease called listeriosis, a bacterial infection which usually results from ingesting moldy feeds or hay. My kids don't get feed and I was pretty certain that my hay wasn't moldy, but I treated with the recommended antibiotics anyway. I lost him. The second one I took to the vet, but he was stumped. He gave him antibiotics but the little fellow died within

24 hours. The third one went down so fast that I barely knew what happened. Needless to say, this turned me against early weaning.

Two months of age is generally considered an acceptable age to wean goats. By this time they are eating solid foods so it is the age at which breeders start selling kids. It is the age at which bucklings are often separated from their mothers because they can become sexually mature that young. It is fully possible that they can impregnate any doe or doeling in heat. From reading various goat forums, however, I learned that many goatherds leave their baby bucks with their mothers until three or four months of age, so I began doing that too. They seem to do so much better if left on their mother's milk for as long as possible (a dam will naturally wean her kids at about five or six months of age). I haven't lost a buckling since I started doing that.

Then something unexpected happened. It was from that incident that I think I finally figured something out.

It was the year Lily had her triplets: a flashy moonspotted buckling I named Splash and two black doelings, one with spots and one without. I named these Dottie and Sissy. They all gained weight well from their mother's milk alone, and I never had to supplement with a bottle. They always rushed her at the same time and with all three of them pushing and shoving and jockeying for a teat, I couldn't really tell who was getting what and just assumed they all got at least some milk.

They were about three months old when, one morning, there was Dottie, weak, wobbly, and alarmingly thin. She was standing alone with a spaced-out look and trembling. What had happened? Wasn't it just yesterday that all the kids were running and leaping through the back gate, down the hill, and into the woods? Or was it the day before? How could this have happened so fast? Why didn't I notice? This was exactly what had happened to the three bucklings I'd lost. How could it be happening again?

The hard part about diagnosing animals is that they cannot give subjective information. They cannot tell you how they feel, where it hurts, or what led up to the problem. Diagnosis is based on objective observations, and anyone who is doing the diagnosing will likely tell you it often feels like taking a shot in the dark. In my research I found a helpful Symptoms Chart over at the Jack and Anita Mauldin Boer Goat website.[1] Based on that chart, my best guess for Dottie was goat polio (polioencephalomalacia or PEM). This is actually caused by a vitamin B1 (thiamine) deficiency, not a virus or bacteria.

How did it happen? Goat polio is more common in young goats than adults, and one of the causes I found listed was "difficult weaning." I had no idea what that meant, but I did know that an abrupt weaning (such as separation) could mean a sudden change in

I have learned to monitor weight gain with triplets and quadruplets.

diet. This is something goat keepers are warned against, because it upsets the microbial balance in the rumen. Besides digesting what the goat has eaten, these digestive micro-organisms build proteins and manufacture the B vitamins. Precautions are always given about changing feeds, adding fresh pasture, etc., because the ruminal bacteria and protozoa need time to adjust for healthy digestion. I knew to make slow adjustments in an adult goat's diet, but it never occurred to me concerning weaning. Could it be that an abrupt early weaning causes a disruption in the manufacture of thiamine so that the young goat develops a deficiency? The symptoms of that deficiency (depression, not eating, weakness, staring off into space, aimless wandering, apparent blindness, muscle tremors) is what is called goat polio. It is treated with massive doses of thiamine injections.

Injectable thiamine is a prescription item so I didn't have any, but I did have veterinary grade over-the-counter injectable vitamin B complex. I followed the directions for using that, giving injections every six hours until symptoms either cleared or the animal died. It took a week's worth of injections but, to my great relief, she finally pulled through.

If I'm right, then this explains why my two-month old bucklings didn't make it. By three months of age, they must be getting more forage than milk, so that the problem didn't occur in bucklings I separated at that age. As an interesting aside, I once sold a pair of wethers to a gentleman who was looking for wethers as pets for his kids. This was the second pair he'd bought, having lost the first two with the same symptoms I'd experienced with the bucklings I'd lost. His were two months old when he bought them.

People deal with bucklings in different ways. I've opted to let them stay with mom and separate at three months, even though it does result in a lot of hollering. Some goat keepers separate and bottle feed all the boys at birth, because the boys are usually stronger and more aggressive than the girls. Bucklings can and will push weaker sisters out of the way. A doe only has two teats so if there are more than two kids there is a risk that one or more of them may not get enough to eat. Pat Showalter of the Kinder Goat Yahoo Group recommended offering a bottle before letting kids back in with their mother after morning milking. The hungriest ones will be eager for the bottle. The more well-fed will be willing to wait for mom.[2]

Another challenge to diagnosis was that the symptoms appear very similar to listeriosis from moldy feed or hay. Also called "circling disease," it requires the immediate use of antibiotics and usually occurs in adults, whereas goat polio most often affects young goats. Dottie's age plus the fact that all the goats had been on pasture only was my best clue. Her response and recovery to the B vitamin injections certainly confirm in my mind that it was goat polio.

Of course I separated Splash immediately and put him in with the big billy boys. He wasn't used to the bottle, but I syringe fed him several medicinal doses of milk over the next several days. Dottie had a few ups and downs but gradually got stronger and started eating again. She was slower to gain weight than her sister, but eventually caught up.

Figuring this out gave me both relief and regret. I was glad to know what happened and how to prevent it, but it saddened me to think that curing those three bucklings had been within my abilities, if only I'd known what to do at the time. It was discouraging that my vet didn't make the correct diagnosis, or I might have saved my little Pygmy buckling, Chipper. It was hard not to think what a difference this might have made toward my success with my Kinder Starter Kit.

Goats are able to synthesize their own B vitamins provided the rumen is healthy. Dietary support is the best prevention, and plant sources of thiamine include kudzu, rosemary, sesame seeds, sunflower seeds, dried sage, and thyme.[3] Even so, injectable vitamin B complex and yeast extract are now something I always keep on hand. I keep a closer eye on kids and am more proactive if they appear weak or disinterested. It won't hurt them and often seems to help.

STEWARDSHIP, BALANCE, AND MARKET MINDSET

"What are you going to do with all of them?" Dan asked. It had been our most prolific kidding season yet, and we had ten little goat kids bouncing around our one acre back pasture.

"Well, I don't have to think about that yet," I replied. "They need to grow up a bit so we can evaluate which ones to keep."

Actually, I'd love to keep all of them, but that isn't possible. As Dan and I talked, I commented how in biblical times one's flocks were a sign of one's wealth. Not so today. To the modern mindset such a notion is primitive and ignorant. After all, wealth is all about money, isn't it? Accumulating wealth is about investments, IRAs, and 401(k)s, right? Not about having more goats. Animals are an expense. In fact, too many animals can seem a liability when one considers how expensive it is to keep them: feed, hay, minerals, fencing, housing, equipment, vaccinations, wormers, vet bills. It all adds up. Unless one is a modern production farmer, goats, indeed most livestock, are more in the hobby category. In other words, instead of making our living from our goats, we enable others to make their living from us as we buy their goods and services in order to keep our goats.

I don't think Dan and I are the only ones who want to break out of that mold. We don't want to simply be another marketing demographic, whereby we can only trade money to meet the needs of our animals. We don't want to have to *go* to work to support our animals. We want our animals to *be* our work and our lifestyle. We want to meet our needs with the land. We want to be more self-sufficient.

Self-sufficiency is a much misunderstood term, I think. When I use it I do not mean isolating ourselves from the world as many assume. Rather, I mean we do not wish to be completely dependent upon the industrialized consumer system to meet all our needs. Especially when that system isn't satisfied with simply meeting human needs, but instead attempts to create them as well. Advertising and trend-setting are designed to support an elaborate and complex system of goods and services which can only be obtained with cash or one of its electronic representatives. Its demand for ever-increasing profits, in fact, necessitates this. We consumers have to continue to buy more, so that they can make more. To Dan and me this system is not sustainable.

Self-sufficiency enables us to be actively involved in the processes of meeting our daily needs. It means taking personal responsibility for those things. There is a sense of purpose and freedom in that. That sense of purpose, especially, is alarmingly missing in today's culture. Money, leisure, and entertainment do not offer fulfillment as many hope. This is evidenced by the increasing social unrest we see, usually disguised as various social and environmental causes.

For us, learning to be more self-sufficient means learning how to find a balance. When it comes to our animals, it means keeping no more animals than our land can support. This is because we have a very strong sense of stewardship regarding both. We want both our land and our animals to flourish. We do not want to sacrifice the land for livestock. We do not want to run it down and wear it out.

Successful self-sufficiency also requires contentment. We must learn to be satisfied with whatever is within our means to produce, rather than always craving more. What Self produces, must be sufficient. That's a matter of mindset and also requires balance. I cannot become so fixed on any one thing that I neglect everything else.

I admit we are still working on finding that balance, and are constantly wondering how many animals we can keep. I'm not just referring to grazing, but to being able to grow what our animals need to be healthy. This is why we are investing in our soil by remineralizing. If we feed the soil we'll feed the plants. The plants will feed our animals and us. It makes so much more sense than trying to meet nutritional needs through purchased feed formulations and vitamin and mineral supplements. Maintaining a balance means determining which goats to keep and which ones to cull. By cull, I don't necessarily mean kill; culling simply means selecting which ones to cut from the herd.

Our winter of 2013 - 2014 held an important lesson because it was the coldest we'd experienced since buying our homestead. Our winter pasture and fall garden went dormant, leaving us to rely more on purchased hay and feed. I realized that besides growing and storing more, it would help to have a smaller winter herd every year. Sales and trades, however, require an interested party with whom to transact. When I tried to sell two of my young goats I had only a few nibbles of interest.

This was an important lesson too. If I don't want to keep too many goats over winter, I need to consider the time of year I offer them for sale, and I need to price them to sell. The expense of feeding them all winter could obviously offset the few extra dollars I might get if I waited to sell them according to the prices set by our local market mindset.

That market mindset is a curious thing. It appears to be based on retail prices whether they are relevant or not. For example, have you ever noticed how thrift store prices go up whenever retail prices go up? That never made sense to me because everything thrift stores sell they

We've learned to be prepared for extremes in weather conditions.

receive free as donations. We could assume it is because of increasing overhead, but changes in utilities, rents, and wages take months or years to catch up with the rising cost of goods.

I first noticed this with the high price of used weaving equipment. I had to wonder why a loom that sold for $1000 brand new, was priced $3000, 20 years later. Hadn't the first owner already gotten their money's worth? We pondered this same thing when we first started pricing used farm tractors. Why is a 50-year-old tractor which cost $1500 brand new, now priced $4000 when it isn't even running? And why does tacking "antique" or "vintage" onto it make it worth more? I know all this makes sense to some folks, but it doesn't to me. To me, the correlation between inflation and used goods is based on an artificial sense of value. It is based on the very economic mindset we see as unsustainable.

Even so, I did not initially apply this to selling goats. I would set prices according to local going rates, choosing something mid-to-low range of what others were asking for their goats. Usually I didn't have trouble selling them until I tried to sell those two. They were healthy, good-looking goats and I didn't think I'd have any trouble selling them. I advertised them on Craigslist and waited. No one was interested enough to actually buy them.

Our goal is to maintain a balance between the health of our critters and the health of the land.

I had to admit to myself that current asking prices for goats were higher than I would have wanted to pay. Having always tried to live my life by the Golden Rule ("Do unto others as you would have them do unto you"), I had to admit that I was being something of a hypocrite by pricing Rosie and Hooper as high as I did. As I contemplated what to do, I realized that if I want to support my goats from our land, it would be better to give them away rather than feed them all winter because of some ethereal notion of their "worth." The bottom line is the health of my land, not making money from selling an occasional goat. In fact, I could easily overuse my land by waiting to make those few extra dollars from that goat. It could easily cost me more to rebuild my soil than what I wanted the goat to be worth.

I offered Rosie for sale again, cutting the asking price by almost 50%. The family that bought her was so excited to get her, and I realized I had forgotten the priceless satisfaction of helping someone else. I used the money to buy hay for the winter. I used the same pricing scheme for Hooper and two others, which were quickly sold as well.

In our quest for self-sufficiency, I realize that we can never truly meet all of our needs by ourselves. Actually, the very nature of the goal necessitates community, but a community of like-minded folks. The modern world says everything is about money, but it didn't used to be that way, and it doesn't have to be that way now. I believe the trend toward modern agrarianism is the movement of folks away from the money model and back toward an economy and social structure based on the land and a sense of community.

The bottom line is being a good steward of what we have. If I keep that in mind, I can keep my priorities clear and straight. I make my mistakes, but this is always what I come back to. It's what always sets things right once again.

KINDERS AT LAST

When Lily gave birth to her beautiful set of spotted triplets, the big question on my mind was who had sired them. Their unknown paternity was the result of the Lily/Gruffy/Hooper/Alphie fiasco as described in "The Game Changers." While it was fun to anticipate developing my own dual-purpose breed, I knew in my heart that I would prefer Kinders. Experimenting was interesting, but I knew it would take years to properly evaluate the result. Kinders were already proven in the traits I wanted. Either breed could help us fulfill our prime directive and I wasn't sure which way to go. So I prayed. If I could find any Nubian/Pygmy-cross goats for sale, then that's the direction I would take, even if they couldn't be official Kinders. I mention "official" because the word "Kinder" is trademarked by the Kinder Goat Breeders Association. Only goats registered with the KGBA can be called Kinders. All other Nubian/Pygmy-cross goats can only be referred to as "Kinder types." However, it wasn't so much the name as the breed characteristics I was after. If that's all I could get, I would be satisfied with that. If I couldn't find any Kinder type goats, I would be content with my Kikobians.

Within a couple of days I found a Nubian/Pygmy buck for sale. I snapped him up. When Lily's kids grew too large to be half-Pygmy kids, I paused again because things weren't working out as I'd hoped. Still, I couldn't help but peruse Craigslist for more Nubian/Pygmies, because what else would I do with my new buck? When I found a doeling, I bought her, after selling off some of my Kikobian kids to pay for her. Then I found two more Nubian/Pygmy-cross doelings. I sent off an email to ask about them and kept looking while I waited for an answer. To my surprise, I found two registered Kinder does for sale, and only about 150 miles away. They were not being sold by a breeder, but by someone who had purchased them from a breeder farther away. These were more expensive, but this is what I had prayed for. Still, it was an emotional dilemma because I had gotten used to accepting an alternative. More prayer.

In the end, Proverbs 31:16 came to mind, "She considers a field and buys it; From her earnings she plants a vineyard." I had a bit of savings from book royalties; what better to do than invest it in Kinders?

I sold the rest of my Kiko, Nigerian Dwarf, and Kikobian goats. I bought the Kinder does, and learned about a Kinder breeder only two states away. From her, I purchased a registered buck for my new does. A second buckling followed soon after that, flying in all the way from California. He seemed like another miracle because it meant finding an airline that would take him plus meeting all the requirements to ship him across the country. At the end of the year I renewed my membership with the Kinder breeders' association, suddenly, but providentially, finding myself very official indeed.

This past spring my two Kinder does gave me six kids: a set of quads and a set of twins. I sold two bucklings and traded a doeling. I kept two doelings and the remaining buckling will be for chevon. In addition to the trade I bought two more registered doelings. My little Kinder herd is growing.

In terms of our prime directive of working toward sustainable self-reliance, the breed of goat is largely irrelevant. All goats provide the basics of milk, meat, kids, manure, and brush control. All require shelter, feed, minerals, water, protection, and whatever is necessary to keep them happy and in top health. All can have health issues and require veterinary care. The kids, of course, mean herd growth, but also having something to barter or sell. One of the things I learned from my Kikobians was that, in general, folks are looking for purebred goats. I always found it easier to sell my purebred goats (with or without papers) than the crossbred ones no matter what I called them. And registered goats were easier to sell than unregistered. As my California buckling's breeder said to me, it requires just as much to

Kinder does are excellent mothers.

raise unregistered goats as it does to raise registered goats. If Dan and I were going to try to make a living from our homestead, it made better sense to produce a quality product that people wanted to buy. As I contemplated that, I felt a sense of closure with my decision making. I had taken the right step.

Some folks will tell you that if you give up a thing it always comes back to you. I cannot say I have necessarily found that to be true, but it certainly was in this case. I have never doubted what Dan and I are doing on our homestead. We have long believed that the agrarian life is the best one. Not necessarily rural agrarianism, but the social and economic model it presents. What will the future hold for my Kinders and me? That will be the stuff of many more goat tales to come.

TOWARD SUSTAINABLE GOAT KEEPING

I've made much of self-sufficiency in the previous chapters. I made much of it in my first book, *5 Acres & A Dream The Book: The Challenges of Establishing a Self-Sufficient Homestead*. I make much of it on my blog, *5 Acres & A Dream The Blog*. All of this ought to beg the question: how well are we doing toward achieving that goal in regard to our goats?

In truth, I would have to say not as well as I wish. If any of the mega disasters forecasted by survivalists and preppers did strike, I would worry about feeding and protecting our goats. In a gentler scenario, such as permaculture goat keeping, we are on the right track; we simply aren't getting there fast enough to suit me. Two elements are key here: the ability to grow a year's worth of what we need and the ability to store it. These two things are contingent on set-up, equipment, knowledge, the right weather, and time. That means there has been a learning curve and much trial and error. We've had both success and failure, and it's tempting to think that our failed experiments were a colossal waste of time. On the other hand, we've learned a lot from our mistakes. I only lament that we don't have an older agrarian generation to teach us the knowledge and skills we need now. What progress have we made, and what have we learned?

PASTURE, FORAGE, AND HAY

When we bought the place, three of our five acres were cleared, but none of it was fenced. Our first fencing project was described in "But First, Fencing" and encompassed about one acre. This is where we put our first goats. The next year we fenced in another one-acre plot, and divided the first for an area to grow corn. The following year we fenced off an area for bucks. Two more years saw two areas of our woods fenced to give the goats woodland browse (and to control the kudzu). We now have enough fenced areas to rotate grazing for our goats, chickens, and pigs, but it took nearly five years because fencing is hard work and expensive.

That first year our best pasture area grew tall in fescue and vetch. Because it wasn't fenced at the time, we cut it for hay. We used a scythe, straight garden rake, and wheelbarrow. We did not have anywhere to store the hay, so we made piles on the ground and covered them with tarps. We lost those first piles because we did not know that hay needed a dry floor and air circulation to keep from growing mold and mildew. It was a valuable learning experience, but we lamented the loss due to our ignorance.

Another thing we did not know is that good pasture requires maintenance. That same best pasture seemed to disappear after a couple of years. The vetch it once grew did not come back and the grass did not look the same. About that time I read *Natural Goat Care* by Pat Coleby and *Hands-On Agronomy* by Neal Kinsey. It was from these that we learned about remineralizing our land as part of our soil building. I detailed this experience in *5 Acres & A Dream The Book*.

We haven't been able to keep our original goal of remineralizing an area per year, but neither have we given up on it. Besides the expense of shipping in the minerals, it requires preparing the ground and working the minerals into the soil. This is necessary because some of them are water soluble and would wash away in the rain. Initially we did not own a tractor and had to rely on neighbors who did. Purchasing a two-wheel, walk-behind tractor was a big step toward being able to manage this task. We also bought pigs and have been amazed that they do a much better a job at turning the soil! In addition, they eat many of the weed roots. Our plan is a four- or five-year rotation, starting with the

Our sickle bar push mower is a favorite piece of equipment for cutting hay.

pigs in the most needy forage area. That will be followed by nourishing the soil and growing field crops. It can then be foraged by our critters and replanted back to pasture grasses, legumes, herbs, even vegetables. What the goats don't eat can be cut and dried for hay. This will be our land stewardship cycle.

Pasture forage includes both grasses and legumes, and we have learned that even perennial forage plants do not last forever. I did not understand this as I fretted over what had happened to our fescue and vetch. I wondered what we had done wrong. I now know that good pasture requires maintenance. From my research I've learned that forage plants can be either annuals or perennials, and are divided into warm and cool weather types. For the southeastern U.S. these are:

COOL WEATHER FORAGE:	WARM WEATHER FORAGE:
annual rye	Bahaigrass
chicory	Bermuda
fescue (low endophyte)	buckwheat
oat	Egyptian wheat
orchardgrass	lespedeza
timothy	millet
various clovers	
vetch	
wheat	
winter peas	

Two others are Johnson grass and kudzu, although both are considered invasive. Both grow here in patches, so I cut and dry them for the hay pile. Fescue, particularly tall fescue, is precautionary. The seed can be infected with an endophyte fungus which causes ill thrift in grazing animals including goats, sheep, horses, and cattle. Endophytes bind copper, selenium, zinc, and cobalt, which result in all the problems associated with mineral deficiencies. Endophyte-free varieties of fescue are available, but apparently they are not as hardy as the endophyte-infected.[1]

Since fescue was the primary grass in our front pasture when we bought the place, that information was of particular interest. Most of Jasmine's problems pointed to copper deficiency, even with a good, free-choice goat mineral available. Looking back I can't help but wonder if the fescue wasn't part of the problem. On internet goat forums and websites there are mixed reports about goats grazing fescue. Many claim their goats do fine. Others have problems. Immediate solutions include vigilant supplementation of selenium and copper, and making sure animals are offered a large variety of browse to decrease the amount of infected fescue consumed. Long term, it means replacing the fescue with other types of forage. Although I

We cut as much of our own hay as possible, although I buy hay to supplement as needed. The cattle panel hay feeder was cheap to make, but it is wasteful.

initially lamented the deterioration of that front pasture, I realized that the disappearance of our fescue grass was a blessing in the long run.

All the forages I listed can be used for hay as well. In a pasture rotation plan, one area can be set aside for growing hay, although the goats usually leave enough to cut in their grazing area. Like all farmers, we try to plan for cutting, drying, and raking in when there is no rain. We have learned how to store it properly so that it stays dry and mold-free. We've invested in equipment too; in addition to the tractor, I purchased a used sickle mower from the earnings from my first book. It is much like a lawn mower but with a sickle bar for cutting long grass close to the ground. Dan still loves his scythe, but we both understand that higher tech tools and equipment can be a huge asset.

One thing that still puzzles me is how to create and maintain a truly sustainable pasture, i.e. one for which I don't have to buy seed every few years. Like all ecological systems, grazing and forage areas change over time. In permaculture this is called succession. This is the "progressive change from one ecosystem or habitat type to another by natural processes of soil and community development and colonization."[2] There are a number of theories as to the whys and wherefores of this, all of which are interesting but largely irrelevant for practical purposes. My seasonal observations serve me better here. For example, every year seedlings of oak, pecan, pine, sweet gum, and cedar trees sprout up, parented by trees on our pasture peripheries. Previously unseen plants will pop up in the middle of the field, from seeds which I presume are dropped by birds flying overhead. Sometimes weather gives an unwanted plant an advantage over others. Just last summer we had a dry spell which gave drought-tolerant ground ivy an advantage over the grasses and legumes we had planted. Also, the goats themselves can eliminate a plant they particularly like to eat. How to maintain good pasture without having to buy seed every so many years is something I have not yet figured out how to do.

How many animals a pasture can support is a common question, but the answers are vague and always prefaced with "it depends." And it does. It depends on the kind and number of animals as well as the kind and quality of the forage. The quality and availability of forage depends on how well the pasture is maintained plus the weather. As every gardener knows, weather is always the unknown factor. It is unpredictable and uncontrollable, yet it is key.

During mild winters our cool weather pasture can supply fresh forage for our goats all winter long. If the winter is severely cold, everything goes dormant and we must rely more on hay. During summer, our rainwater catchment has been a garden saver. For our larger forage areas, we have to hope and pray for rain. We've tried to learn when our annual dry spells are and to plant accordingly. We try to plant so that the seed won't have to sit on the ground waiting for rain, or so that newly sprouted seeds don't wither and die from lack of it. This affects not only pasture areas but field crops as well.

GRAINS AND LEGUMES

In the beginning, I assumed we would have to grow grain for our goats. Grains are high in energy but low in crude fiber, as are the beans or peas which are added for protein. Referred to as "concentrates," these are considered the necessary non-forage portion of the diet. The modern pelleted form usually contains corn and oil or fat for energy, soy or cottonseed meal for protein, plus added fiber, vitamins, and minerals for a complete feed in a bag.

With Dan's and my goal of feeding our animals from our land, I assumed I would need to learn how to make my own concentrates. I would need to know what grows well in our area, plus I wanted to find a soy substitute, because soy requires extra processing to make it digestible. To grow a year's supply of these things would be more challenging than the garden. It would require good soil and adequate moisture, but at a scale needing more land, more water, larger equipment, and more storage space. Because of that I wanted to do my homework first.

As I researched what to grow for our goats, I learned there are a wide variety of opinions about what to feed them. Some folks are able to maintain healthy, productive goats without grain. Others assume it's a must. They believe grain is essential to a goat's health and well-being.

The reason for feeding grain and legumes to ruminants is to enable them to reach their "genetic potential" for maximum milk production and rate-of-gain, i.e. weight gain for meat.[3] One of the questions I had to ask myself is, do I really need maximum production? If I want to make any significant income from my animals, then my answer will likely be yes. As someone who simply wishes to live on what I can grow and produce for myself, then I need to consider my answer. Is it possible to be satisfied with what my goats can provide without being pushed to their limits? Can I adapt our diet to true seasonal eating? More importantly, does our feeding regimen fit our values of stewardship and self-sufficiency? Is it in my goats' best interests to be pushed to maximum production or will I burn them out? Is it possible for my goats to thrive without a concentrated type goat feed?

In *Alternative Treatments for Ruminant Animals*, Dr. Paul Dettloff, DVM comments that his veterinary textbooks of the 1950's make no mention of acidosis. It was not a problem when ruminants

Good quality forage is a must.

(cattle in his example) were fed on hay and pasture alone. The problem developed when the standard feed for cattle became grain, particularly corn, both as feed grain and as silage. The result has been acidosis, and its various symptoms are now common: hoof problems, loose and runny manure, decreased resistance to disease, decreased butterfat, and shortened lifespan. Autopsy reveals an enlarged, yellowing liver.[4]

In goats, symptoms of acidosis are similar to those seen in cattle including indigestion, decreased appetite, dehydration, depression, weight loss, foot problems, scours, B vitamin deficiencies, decreased resistance to disease, and eventually death. Preventative measures include adequate roughage (especially long-stemmed grass hay), whole rather than cracked grain, feeding hay first, and slow changes in diet to give the bacteria in the rumen time to adjust.[5] Acidosis is also the reason goat owners leave out free-choice baking soda, so the goats can self-treat mild cases.

The problem is that the rumen is not designed to digest grain and seeds. It is designed to digest roughage, which is a generic term for the long-stemmed, high fiber plants which make up forage, browse, and hay. According to the classic *Feeds and Feeding* by F. B. Morrison, the rumen is unique because it contains digestive bacteria which are able to break down plant cell walls, particularly cellulose and pentosans. This is something digestive enzymes cannot do, which is why roughage is low in digestible nutrients for monogastric (single stomach) animals. Ruminants, on the other hand, are able to assimilate more nutrients from roughages. The longer the roughage stays in their rumen, the more nutrients they can extract. However, the digestive bacteria will also digest the starches and sugars found in grain (and in molasses, which is used as a binder in processed feed pellets).[6] Dr. Dettloff, writing sixty years later, makes the connection with acidosis. The digestion of starch throws off hydrogen ions which lower pH, initially in the rumen but, eventually, systemically. The body's cell membrane's sodium-potassium pump becomes unbalanced so that the immune system suffers and the animal's condition deteriorates.[7]

What goats need are plant materials with long coarse fibers. These stimulate the rumen to function properly in what is known as the

Good quality hay is necessary for the roughage effect. We like this double-sided hay feeder because it lets more goats eat without being chased off by the others.

"roughage effect." During digestion, plant matter is broken down and begins to ferment. Because it is not very digestible, roughage requires re-chewing to further break down the cell walls to release nutrients. We know this as chewing the cud. This action also neutralizes rumen pH because goat saliva contains a buffer.[8] This buffer is a naturally produced bicarbonate, which according to M. Hadjipanayiotou, is apparently superior to baking soda (sodium bicarbonate) in its ability to regulate rumen pH.[9] The more the animal chews its cud, the more bicarbonate is released into the digestive system. Grinding roughage into small particles (as for making pellets) greatly reduces this effect and the rumen's ability to digest and buffer properly.[10]

Something that surprised me is that goats don't need high energy feed (grains) to keep warm during cold weather. Among other nutrients, ruminal fermentation produces volatile fatty acids (VFAs). These provide over 70% of the needed energy supply,[11] enough to enable the animal to stay warm.[12] Robert L. Johnson of the International Dairy Goat Registry concurs, "They need plenty of real roughages–tree bark, dry leaves, poor-quality hay, even straw. If you give a goat a big bowl full of high-protein feeds on a cold winter night, you are actually chilling the goat; still more energy is needed to digest the meal, and goats can get pneumonia."[13]

Keenan Bishop summarized it well for Kentucky's *The State Journal,* "Goats do not perform well on high concentrate or high starch diets; however, they perform best on feeds in the range of 55-70 percent total digestible nutrients with just enough nitrogen in the gut to produce microbial protein and enough digestible fiber to produce fatty acids."[14]

The subject became more complex from there. In addition to crude fiber (CF), some articles now discuss acid detergent fiber (ADF). Basically, higher ADF means lower digestible energy and is used as an indicator of when to add grain to the diet. When I could not readily find charts with this information, I began to wonder how complicated I really wanted to get. Isn't the simple life supposed to be, well, simple? My conclusion was that everyone recognizes the inability of ruminants to digest high starch concentrates, but their solutions are finding ways to treat symptoms rather than eliminate the cause of the problem. They add rumen buffers, neutralizing agents, and rumen modifiers to concentrates, trying to obtain a profitable balance between production and health. Exact, scientifically-calculated ratios for roughage, grain, and additives may make sense to those in a feedlot situation, but it did not seem the best plan for our goals and our goats. To me, that seemed like a game of Russian roulette that I don't want to play. Free-range goats will choose a large variety of things to eat if available, and they are healthier for it. Our job is to steward our land to keep that variety available.

That still left the question of protein, another much talked about topic in regards to feeding goats, and another reason why folks feed concentrates. I've written many a blog post reporting my research on the protein content of things we can conceivably grow on our homestead: wheat, oats, amaranth, black oil sunflower seeds, grain sorghum, corn, cowpeas, comfrey, etc. How to make a homegrown feed mix containing the recommended 16% crude protein (CP) was a concern, because the things we can grow don't contain enough CP to get that percentage. That's why soy is commonly used in commercial feeds.

It was when I read Pat Coleby's *Natural Goat Care* that I learned that overfeeding protein can actually cause health problems such as mastitis, acetonemia, milk fever, ketosis, foot problems, and also mineral deprivation, especially copper. In addition, legumes which are used as protein sources (soy, alfalfa, beans, peas, tagasaste, clovers, etc.), are goitrogenic, which means they interfere with the thyroid's uptake of iodine. Fed in excess, they deplete iodine and can cause thyroid problems. I know that when I feed my goats lots of alfalfa, they consume more kelp, which is their source of iodine.[15]

Another concern with feeding large amounts of legumes is that they contain phytoestrogens (plant based estrogens). These can affect ovarian function, fertility, and milk production.[16] Also they can be a contributing factor in ovarian cysts.[17] These things raised red flags in my mind in regards to whether or not the protein they provide is worth the risk. To put it another way, do goats really need all that protein?

Most dairy goat feeds contain a minimum of 16% protein (some higher). Proponents of natural goat care recommend 12 to 14%.[18,19] According to the online *Merck Veterinary Manual,* the very minimum amount of crude protein needed by a non-working goat (i.e. not pregnant, lactating, or in rut) is 7% crude protein.[20] These figures are for crude protein (CP, which is actually nitrogen content). Of that, roughly 70% is digestible.[21] That means 16% CP yields approximately 11% digestible protein, 7% CP would be about 5% digestible. That seemed shockingly low to me after years of trying to figure out how to make my own rations containing 16%. In light of the other information, however, I was beginning to reconsider.

An extremely helpful article was one I found by Dr. Robert J. Van Saun DVM, of Penn State University. In "Dairy Goat Nutrition: Feeding for Two (How to properly feed the goat and her rumen)," he states that milk production can be increased by decreasing grain and maintaining a high roughage diet. How is that possible? Because the microbes in the rumen produce the building blocks of protein through bacterial fermentation.

"The dairy goat derives a majority of her energy and protein from microbial end products or the microbes themselves. Bacteria contain approximately 60% protein, which is of high quality and digestibility. In other words, the more we make the bugs grow in the rumen system, the less additional more expensive feedstuffs we need to provide the doe."[21]

He goes on to state that in dairy cattle, microbes can provide protein for up to 50 pounds of milk.

According to another source, ammonia builds up in the rumen when rumen degradable protein exceeds the capacity of the rumen microbes to assimilate it. The ammonia is absorbed into the blood and converted to urea in the liver. This conversion process takes energy that could be used for making milk. This is why too much protein in the diet decreases milk production. It can further create a negative energy balance which can eventually result in reduced fertility.[22]

All of this supports a low grain, high roughage diet. Grain can cause problems, but roughage keeps the rumen active, healthy, and able to extract those protein building blocks.[23]

What does all of this mean to me, in terms of growing grains and legumes for our goats? It means that if I can provide high quality forage, both fresh as pasture and browse, and dry as hay, then I do not need to focus on growing grains and legumes for them. I can put that land to use growing winter greens and root crops to supplement their diet rather than buying concentrates. If I do feed them grain and legumes, I would rather include it in their hay as whole plants, i.e. before the wheat has been threshed and with the peas still in the pods on the vine. The goats can get the nutritional boost from the wheat and peas, but with the buffering, roughage effect of the stems, pods, chaff, and leaves. They eat it all and the chickens happily clean up any dropped seeds.

MINERALS

One other important area of goat nutrition is minerals. Goats seem to have a high need for minerals, which is evidenced by ongoing discussions amongst goat owners regarding mineral deficiencies and their symptoms. I'm going to say it is why goats prefer to eat what is known as browse. Browse is a broad term for the leaves and tender stems of bushes and trees, also weeds. Unlike short-rooted grasses, all of these have deep roots which pull up minerals from deep in the soil. Give a goat a choice between a nice pasture or a weedy, overgrown lot,

Minerals should be available to both adults and kids on a free-choice basis. The concrete block lets kids and smaller goats access the minerals easily.

and they will choose the weedy lot. This is why goats make an excellent choice for vegetation management.

Eventually, however, they eat all that unwanted brush and end up on a diet of pasture, hay, and grain. Then, the most common way to meet goats' mineral needs is with a loose mineral supplement. Mineral salt blocks are also available, but free-choice loose minerals are preferred by those who raise goats. Additional minerals are also added to goats' diets as symptoms warrant. The most common deficiencies are copper, selenium, and cobalt. These stem from regional soil deficiencies, and must be addressed on an individual basis. Commonly, this is done by adding specific mineral supplements to the goats' feeding routine.

Previously I mentioned having our soil tested for micronutrients and then purchasing these nutrients and adding them to the soil to remineralize it. Some folks think this is unnecessary and that by simply adding manure and compost, the soil will be taken care of. Plants grown on mineral deficient soils, however, will be mineral deficient as well. Being consumed and digested by animals which also have mineral deficiencies will not add those missing minerals to the manure. That is why I believe remineralization is the best plan. The initial investment is high, but over the years a savings is seen in the health of the land and the animals.

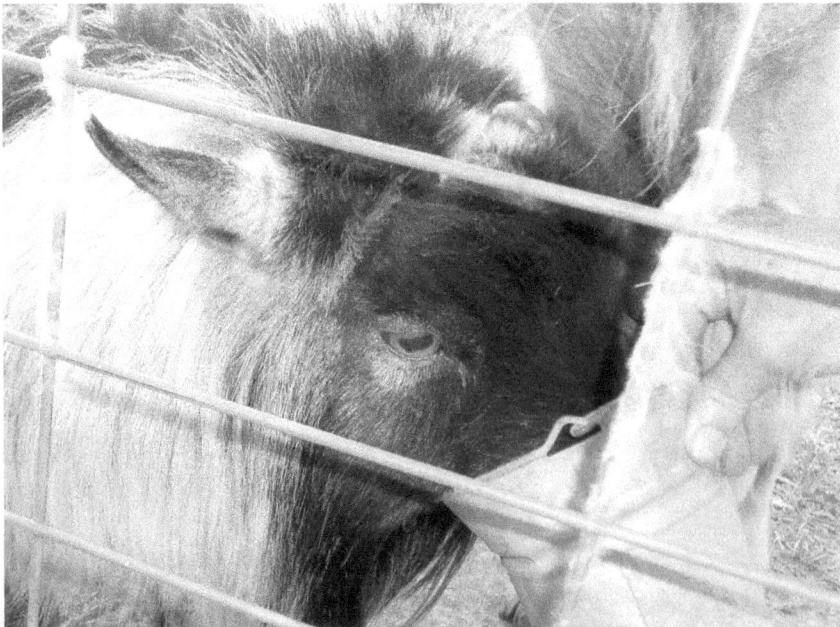

Bucks need their minerals too, either sprinkled on their feed or offered in mineral feeders and other small containers.

Some plants are especially rich in specific vitamins and minerals, and are another valuable resource for goat nutrition. For example, beet greens and chickweed are rich in beta carotene, chicory and dandelion are good sources of copper, and comfrey and amaranth leaves contain good amounts of calcium. These can be grown in goat forage areas, or collected and dehydrated to feed later. See "Resources" for where to find more information.

PUTTING IT ALL TOGETHER. NOW WHAT?

At some point I had to extract myself from the never-ending trail of the whys, whats, and wherefores of feeding goats and ask myself how to put this information to good use—how to evaluate what we've done so far, and decide where we need to focus our efforts in the future.

One thing my research did was to confirm our original goal of focusing on soil and plant health. It also reinforced our goal to grow as much of our own feed as possible. Hay and feedstuffs grown on others' poor soil won't help me maintain top health in my goats. Also it was a relief to know that we did not need to grow a lot of grain and legumes specifically for the goats. We should focus on our paddocks for forage and hay, with some sort of supplemental feed to make up for any dietary shortfalls.

In regards to forage and hay, we are fortunate that our climate is conducive to growing cool weather grasses, legumes, and vegetables most winters. Even so, I will need hay available for extended rainy periods or when it's cold enough for our pastures to go dormant. One concern I've had about our hay quality has been timely cutting for the most nutrition, i.e. in the leaf stage before it goes to seed. Sometimes we're able to cut it then, sometimes weather or more urgent needs cause delay, so that the hay has more stems and seeds by the time we actually get it in. I now know that the older grasses are also important for the roughage effect. They might not be optimal nutritionally, but they still have digestive value and there is no waste.

Our hay includes more than grass and legumes. I've learned there are a lot of safe edibles for goats growing on our property, and I harvest quite a few of them to dry and add to the hay pile: kudzu, morning glory vines, honeysuckle, seedling trees, also grasses and weeds I trim along fence lines with my hand sickle. I lay these out on a tarp to dry and then toss them onto the hay. I've done the same with sweet potato, bean, and pea vines after harvesting, as well as thinnings from corn, cultivated amaranth, and sorghum. Also I save corn husks. Goats love dried leaves too, and I like to gather these after our first hard freeze. Many of the leaves "freeze dry" and fall to the ground dry but green.

As I've decreased grain I've increased garden vegetables, herbs, and fruits, both fresh and dried. In summer I feed greens such as Swiss chard, amaranth, and comfrey. Also melon rind, overripe or underripe blueberries, trimmings from canning and cooking, overgrown okra, squash, and cucumber. Summer is when I dehydrate and store as much as I can to feed during winter: greens, herbs, blueberries, rose hips and citrus peels from my Meyer lemons. These are used to top dress their feed for additional vitamins and minerals. My homegrown grain is mostly amaranth, but also wheat fed as whole plant hay. Cowpeas are harvested and dried, and added to their feed still in the pods. If I have homegrown flax seed, I give it to them in the pods because they'll eat those too. Homegrown sunflower seed heads can be chopped and fed; no need to separate the seeds from the heads.

In winter they get cool weather greens such as kale, turnip, radish, cabbage, broccoli, and collards. Chopped root crops include sweet potato, Jerusalem artichokes, beets, sugar beets (non-GMO), turnips, mangels, and carrots; also chopped winter squash and pumpkin if I have it. In autumn I collect white oak acorns to add to their feed.

Transitioning to a 100% homegrown diet is slow, and I supplement with purchased feeds as needed. I make my own grain mix of four parts grain (wheat, oats, and barley when I can get it) to one part black oil sunflower seeds. Of this, I give less than a pound per day. I used to feed alfalfa pellets until I learned that they were made from genetically modified alfalfa. Now if I need some filler for their feed, I use Chaffhaye, a naturally fermented, bagged form of alfalfa hay.

After field corn harvest, the goats are allowed to feast on the leaves.

I offer a loose goat mineral, but also top dress their feed with the the mineral lick on page 90 of Pat Coleby's *Natural Goat Care*. It contains powdered sulfur, copper sulfate, kelp meal, and dolomite, although I use dolomitic limestone because it's what I can find locally. (Note that it must be dolomitic limestone, not calcitic limestone, to get the correct calcium-to-magnesium ratio.) Each goat gets a tablespoon sprinkled on their feed. All goats have access to free-choice kelp. I used to offer baking soda, but they stopped taking it when I cut back on grain.

Feed amounts are not strict, but are based on each goat's condition. A doe that looks thin gets more. Some gain weight more readily and don't need as much.

My bucks do well on pasture and forage alone in summer. During rut or when their condition warrants it, they get chopped garden produce (fresh or dried), a bit of Chaffhaye, and the mineral supplements. Kids get a handful of Chaffhaye for the probiotics with a sprinkle of the supplements for vitamins and minerals; just enough to keep them occupied while their mothers eat their portions.

On Wednesday, I add a couple teaspoons of homemade herbal wormer. I will use chemical wormers as needed, but considering that chemical resistant parasites are developing, I use natural wormers to deter and help control them. My wormer includes black walnut hulls, fennel, thyme, and wormwood (except for pregnant does, who get the same mix without the wormwood). I top this off with a clove of fresh garlic, another natural wormer.

Amounts for all goats can vary depending on which paddock they're in. The ones in the remineralized areas seem to need less, the ones in poorer areas get more. My overall guide is our goats themselves.

What I assess, however, is condition not production. By that I mean I am not as concerned with maximum milk production or additional weight gain for meat as I am with their overall condition and health. I keep an eye on their coats—they should be smooth, soft, and shiny. Their bellies should be rounded, not flat or sunken. I feel their ribs and hips to make sure they are not sharply bony. The Kinders, particularly, should have a beefier look than Nubians, with good padding and no sunken flanks. All goats' droppings should be firm "berries," not clumped. I check the inside of their eyelids, looking for a watermelon pink color. Anything paler indicates an increasing worm load. If any goat is not alert and interested in their surroundings, I begin to suspect that something is out of balance. Their health is my bottom line. This approach means that I must adapt our diet to what our goats produce rather than try to push production to meet a set demand.

FUTURE PLANS

Besides continuing to improve soil and plant health in our forage areas, Dan and I have been discussing ways to increase plant diversity, especially herbs, vegetables, and the leafy shrubs and trees considered browse. By watching our goats eat, we know that they will go for the greatest variety available, taking a bite of this and a bite of that. Also we've learned that goats can pretty much wipe out things they really like. Pasture rotation helps, but some things are slower growing, such as shrubs and bushes. We want to be able to offer these in a way that protects them too.

My first experiment with this was a "fence pocket" made by attaching both ends of a curved cattle panel to an existing fence. In the protected "pocket" I planted Jerusalem artichokes. The goats love the leaves but will eat them down and kill the plants if allowed. The idea is that the cattle panel will let them eat some of the leaves without demolishing the plants. So far it seems to have worked.

We expanded on this idea with our first forest garden hedgerow. We used two parallel rows of cattle panels, about four to six feet apart, to divide our one-acre front pasture into two areas with a strip of land down the middle. The protected area between the cattle panels is where we are planting a permaculture hedgerow for not only the goats, but also the chickens, pigs, and us. Fruit and nut trees go in first, then shrubs, vines, and bushes, along with greens, herbs, and other vegetables. Eventually, we hope to have these hedgerows between all of our pasture areas.

I've begun to consider other things I can use in my homemade goat feed, taking a cue from the fiber fillers manufacturers use—ground corn cobs and corn stalks, for example. My idea was to do this with a hammer mill. In an agricultural setting it is used to make cracked grain feeds. An increasingly common use is to shred materials for making fuel pellets for pellet stoves. It was at a pellet stove website that I found do-it-yourself hammer mill plans and bought a copy. When we had trouble finding everything we needed to make one, Dan came up with another idea.

One of our early purchases for the homestead was a "Yard Machine" to make wood chip mulch from our many sticks and branches. As a chipper/shredder it was pretty disappointing, because it scattered pulverized mulch all over the place, like way across the yard. As a piece of junk to clutter up the carport it did great, and we often pondered what to do with it. We would crank it up every now and then, but it could only handle small sticks and leaves so it was only semi-useful.

Cattle panels along fencelines create pocket areas for growing things like Jerusalem artichokes. The goats can eat the leaves, but cannot kill the plants.

An old gas-powered mulch maker, a 30-gallon metal drum, PVC pipe, and rain gutter parts were all that were needed to make a goat feed chopper. The air outlet on top of the drum is covered with a fine mesh. Best of all, it works!

After discussing the hammer mill, Dan dug out the Yard Machine plus some PVC pipe, rain gutter parts, duct tape, and a 30-gallon lidded drum. The result was our prototype goat chow maker. We ran corn cobs and amaranth stalks through it as our first experiment. The results were not as consistent as we hoped but not bad for a start. The real test was whether or not the goats would eat it! I took a couple handfuls of the finer stuff and mixed it in with the evening feed ration for the girls at milking time. Daphne ate it all and Helen left only the largest chunks of corn cob. Dan went on to do some tweaking, but we definitely called the experiment a success.

CONCLUSION

The old adage that we feed the rumen, not the goat, is so true. I'd not disbelieved it, but until my research I hadn't understood how vitally true it is. The beauty of this is that it simplifies not only what to feed them, but how to provide it. We are on the right track, we just have to continue working toward our goal.

Llama Tales

IN WHICH WE GET A LLAMA

I never planned to get a llama. When I thought about the homestead animals I would get, it was goats for milk and Shetland sheep or alpacas for fiber. But one day I was perusing Craigslist and found a "clearance sale" on weanling llamas, $300 each. Considering that they usually cost thousands of dollars, it was something to think about. Of course I couldn't help but mention this to Dan, although I didn't expect any encouragement whatsoever.

Dan and I have a personal joke that it doesn't hurt to take a look. It's a joke because this is what he said before he came home with a Suzuki 750 motorcycle. That was many years ago but we always grin at the saying. I concluded my casual mention of the llama clearance sale with that phrase, and to my surprise he said, "Well, we either need a guard dog or a guard llama." So off we went with $300 to take that look.

The llama folks had about eight or ten young llamas for sale. They were different sizes and colors, but all were being called "weanlings." This term indicates that they had been weaned from their mothers' milk but are not yet a year old. This opportunity had arisen so quickly that I did not have time to do my usual careful research on the subject. I knew that llamas are herd animals that produce fiber and can be used as pack animals or livestock guardians. I knew very little about their temperaments, personalities, preferences, natural instincts, and care. Ordinarily I would have thoroughly and extensively researched everything there is to know about llamas. This time I did not and had to rely on the llama breeder's expertise.

For some reason I chose the second young llama they showed me. Dan told me he was surprised that I didn't want to look at them all, that I was so quick to say I'd take this little guy. I'm usually not that impulsive, but on that day I was. I gave her the money, she gave me his paperwork, and we loaded him up in the back of our Jeep and headed for home.

The paperwork consisted of a sales receipt and an application to apply for registry if I wished. I wasn't interested in becoming a breeder of llamas, actually I only wanted the one, but I learned a few things from reading it over. First, that his official name was Chocolate Drop. Second, from his date of birth and counting on my fingers, he was actually only five months old.

What did the other critters think? The chickens were fascinated but the goats kept running away from him. Not being experienced with llamas, I couldn't read his reaction to them or his new home, but I was delighted to have him.

CHARLIE

Dan and I agreed that "Chocolate Drop" didn't quite suit. For one thing it was a mouthful, and for another, I wouldn't want the neighbors to hear me hollering "Come, Chocolate Drop." I decided to call my young llama Charlie.

It took Charlie several weeks to start keeping company with the goats. Gradually I saw him grazing with them more frequently. Eventually he started spending the night in the goat shed with them instead of under the maple tree where he'd created a lovely flat spot for rolling in the dirt.

Shortly after Charlie arrived, I bought my second registered Nubian doe, Jasmine. She, of course, didn't know the feeding routine, which was to feed CryBaby and Surprise, my other two does, from a large pan, and Charlie from a small, handheld enamel sauce pan. I had a small feeding pan for Jasmine too, but she wanted to chase the others from the big pan and eat it all herself. That's when Charlie stepped in. When he saw Jasmine try to run off the other two, he got right in there. He shouldered Jasmine out of the way and blocked her so that he,

Llamas like to take dust baths.

Surprise, and CryBaby all ate from the big pan. Of course, Jasmine was upset and the whole neighborhood knew all about it! We soon got it all straightened out and over the next few days things settled down. Fortunately, there are rarely hard feelings in the animal world, and Charlie soon proved to be quite a friend to Jasmine.

One morning, Jasmine didn't go out with the other goats. She just moped around the goat stall, and I knew something was wrong. Goats are gregarious by nature and a goat choosing isolation is not a good sign. The only symptoms she had were teeth grinding (often a sign of pain) and depression. At least, she seemed depressed to me. She had been talkative and friendly when she first got here, but on that day, Jasmine wasn't saying much and wasn't interested in going anywhere. She had no fever and her appetite was still good. In looking her over I couldn't find any injury.

The interesting thing was that Charlie sensed it and did not go out with the others that morning. He stayed right there in the stall with her. For the rest of that day and part of the next, he didn't leave her side. That night, he slept with her and I found them "cuddled" together in the morning. By afternoon, I had talked with the vet and gone to pick up an antibiotic and pain killer. Charlie hung around until after I had given her the injections, and then he went off to graze. The next morning he was out and about before the goats, and Jasmine went out with the other goats too, obviously feeling better.

Learning About Llamas

Llamas are graceful, elegant creatures with an aristocratic countenance. Inquisitive, intelligent, and aloof, they are unlike any animal I'd ever had. They are members of the camelid family and native to South America, where they are used extensively as pack animals in the Andean mountains. They are also used as a source of meat.

One of the questions folks would always ask is, "Do they spit?" While they can spit, I never saw Charlie do it. I did learn to keep an eye on his ears. Perked forward meant he was interested in something, laid back meant he was perturbed.

As herd animals, llamas don't do well alone. Of course, they prefer the company of other llamas, but when kept with sheep or goats, they bond and become naturally protective of them. That's why they are said to make good guard animals. Their care is certainly easier than a livestock guardian dog, plus they are fiber producers, which was much of the appeal for me.

Charlie's first month on the homestead was a learning time for both of us: llama research for me, and trust training for him. Llamas don't like to be touched and they don't like to be caught. A catching

One thing I've learned about feeding critters is that in order for each one to get their allotted feed, they all need to be separated.

pen is considered a requirement, and we made one with corral panels. Charlie was extremely suspicious of it, and try as we might, we never were able to herd him into that pen. So I adopted a different approach.

I started by seeking him out to talk to him several times a day. After brushing "the girls" in the morning, we would all head out to find him. I developed a sing-song "llama, llama" call, adding "Charlie llama," when I found him. All my movements were relaxed and slow. I respected his "humans are not allowed to touch" preference and kept my respectful distance of at least one arm-length from him while I talked to him. The goats all wanted the attention to themselves, so I would let Charlie watch me petting them.

Feeding time for the goats was morning and evening. When Charlie began showing up then too, I started giving him a teeny bit of sweet feed in an old white enamel sauce pan. I would hold the pan out, but wouldn't try to touch him. I'd talk to him while he nibbled his feed, and he'd give my hand a good sniff when he was done, sometimes making contact. Once that routine was established, I started working on touching him. I told him, "touch" and would slowly move my hand toward his neck. We gradually got to where he would let me touch him briefly. That's when I figured it was time to start regular training.

I've always enjoyed hiking and really wanted to train Charlie as a pack llama. He would need to learn to walk on a lead and carry weight on his back. I thought it would be nice to walk him down to the nearby feed store for sacks of feed. First, however, I needed to be able to catch and halter him.

I read that even well-trained llamas have a "Don't catch me" ritual before allowing themselves to be caught. Because of that I always expected it and remained persistent. Being calm and moving slowly are key. Fortunately Charlie had been broken to the halter by his breeder, so he learned to accept my haltering him fairly quickly.

It took one session to teach him to be led. After that we started going out on walks together. Llamas love new things to observe, so we often took a new route. He would follow willingly, but we had to work on going until I was ready to stop, as opposed to stopping and eating when he felt like it. There were many new things to get used to, and he was often uncertain in new situations, sometimes breaking his own no-touch rule. He would stand right up against me as we stopped to take in something new. If he got skittish, we'd stop to examine whatever he was concerned about, and then repeat passing the object or area several more times. I continued to limit touching and petting, and it finally got to where he didn't mind when I massaged his neck and back.

We did have a problem, however, at feeding time. I fed Charlie and the goats together because they lived together. The goats would tend to rush to eat Charlie's small ration first, pushing him out of the way before they'd finished their own. Consequently he would gulp his feed, sometimes choking on it by trying to swallow it whole. I tried a number of feeding arrangements, but eventually had to separate the goats by putting them into the goat shed, two in each stall. I fed Charlie by himself outside. This was a break-through point in our relationship. After he finished he would come over to me and let me scratch his neck. I can't tell you how honored I felt; being allowed to touch a llama is a privilege not to be taken for granted. It created a bond I'd never before felt with an animal. It made Charlie very special indeed.

A Worrisome Turn of Events

We knew something was wrong because even though Charlie's appetite was good, he began to lose weight. He had battled a couple of cases of diarrhea, for which I elicited the help of our veterinarian. The cause always turned out to be parasites and the prescribed wormer took care of it. That could have interfered with weight gain, but even when everything was normal, he continued to lose weight.

My small animal vet was only marginally knowledgeable when it came to llamas, so I turned to the internet to find the expertise and answers I needed. I found an online llama forum and asked about feed and weight loss. I had been feeding him a small amount of sweet feed, which is what his breeder used, but I switched to a llama ration after talking with other llama owners. I continued with his probiotics but couldn't see that the new feeding regimen made a difference. I started feeding him twice a day, also offering carrots and apples. He wasn't interested in these and continued to get thinner.

We discontinued our walks as he became weaker. I would still take him out for fresher grazing in the back yard, which he seemed to enjoy, but it didn't help. Eventually he seemed to stop caring and wouldn't get up. It was at that point that I began to wonder if llamas could have such a thing as "failure to thrive."

I started to research and discovered that, unfortunately, Failure To Thrive (FTT) is common in cria and weanling llamas and alpacas. Although they appear normal at an early age, they later stop growing. Such was the case with Charlie. He never grew the entire time we had him. Dan would sometimes mention that to me, but I assumed it was slow growth rate; llamas aren't considered full grown until they are about four years old.

Pinpointing the cause of FTT is a slow, trial-and-error endeavor, often impossible because the causes are so varied, for example: parasites, rickets, digestive abnormalities, iron deficiency anemia, juvenile llama immunodeficiency syndrome (JLIDS), low birth weight, birth defects, heart defects, thyroid problems, coccidia, being weaned too young, bovine viral diarrhea virus (BVD virus or BVDV), even genetics.[1] Too often, by the time FTT is identified, it's too late.

In Charlie's case, he had been weaned too young. He was only five months old when I bought him, but I later learned that the recommended age for weaning is six months. He was hungry and eating, but what his body needed was his mother's milk. I could only guess that his digestive system wasn't mature enough to obtain the nourishment he needed from grass, hay, and grains alone. It wasn't capable of absorbing much needed nutrients, and the probiotics weren't enough to make a difference. Unfortunately, this was something I could do nothing about. The prognosis was grim, and I knew I had to brace myself for the worst.

LOSING CHARLIE

Now that I knew what was going on, the only thing left to do was to wait. The hardest thing of all is to do nothing. It was hard to let him go, it was hard not to take it as a personal defeat.

On his last morning he couldn't lift up his head. I had already accepted that he was dying, but his death was no less devastating. When it finally came, I was flooded with grief, but I also felt that a burden had been lifted. Dan was at work, but we spoke by phone and agreed on a burial place in our woods.

I clipped Charlie's fleece, feeling that this gave some small meaning to his short life of eleven months. After that, the only thing I felt like doing was digging. I didn't think I could dig a burial hole deep enough, but I figured I could start. I figured I'd dig as long as I could, and Dan could finish it when he got home.

I worked slowly and thought about Charlie. I had become very attached to him because I had worked so closely with him for training. I thought about what happened to him and what I could have done differently. It was hard not to feel responsible. It was hard to accept.

I kept on digging. The ground was wet so the clay soil was heavy, but with a weather forecast of more rain and snow, I kept on digging. I needed a physical outlet for my grief, a channel for my emotions. By the time the sun was low in the sky, I had dug a huge deep hole, big enough for Charlie's final resting place. I laid him in it. I filled in the grave and then covered it with logs and branches, making a large neat pile. This was important to prevent some hungry critter from coming along later and digging up the body. I headed back to the house as the sun was beginning to set.

A LLAMA IN MY FUTURE?

Five years after Charlie's death, we remain llamaless. Initially the loss was too personal to consider a replacement. Especially with animals, we tend to take to heart the hard lessons learned. When folks would ask if we were going to get another, I would say not unless someone mysteriously left an abandoned llama on our doorstep.

Now I sometimes find myself looking around for llamas for sale. Any time Dan says that we need a big animal on the place (he's thinking horse or cow), I immediately say, "Llama." Realistically, the chances of actually getting another llama seem slim to none. Still, of all the critters we've had, I think my llama was my favorite. Is there a llama in my future? Only time will tell.

Puppy Tales

We Decide to Get a Dog

We moved to our homestead with two cats, but often discussed getting a dog. Partnering with a dog on a farm or homestead only makes sense. A dog provides a unique sense of security and well-being around a place. With livestock, we felt we needed some protection for them, a livestock guardian. We'd been told that at one time there had been coyotes, bobcats, and black bears in the area, but no one had seen any of these in over a decade. Several years after we moved to our homestead, however, coyotes had once again been spotted. Dan himself had seen them. Other predators were stray dogs, hawks, and deer, if one considers how deer prey on the garden. My llama Charlie was intended to be our livestock guardian, if he hadn't had such a short life. We had yet to decide what to do when something happened that prompted us to action.

One day I stepped out onto the back porch and my attention was immediately drawn to a chicken fuss. I looked, and saw something running around in the buck pasture. Bushes obscured my view, but I could see it was brown and assumed that it must be our wether, Billy. I couldn't help but wonder why he was running like that. Then I saw it dash off in the other direction. Of course I had to investigate and quickly realized that it was, in fact, a big brown dog. It was chasing panicked chickens. The bucks were trying to hide in the undergrowth of trees.

I took off running, wondering what I could grab to fend it off, throw at it, or use to clobber the living daylights out of it. Dan keeps a rifle and a handgun in the house, but I've only had lessons once and, at that moment, regretted not insisting on regular practice. I definitely would have shot that dog – you livestock owners will understand. Our livestock have just as much right to protection and safety as pets have. A dog is a predator by nature and instinct, and I do not think it should be allowed to maim and kill other animals for fun. I grabbed an old bed rail that was in use to hold down a pile of tarps.

I was fighting mad by the time I got there. When the dog saw me, it ran to the opposite corner of the fence. The goats came to me when I called and I let them out the back gate. As I stomped toward the front, hefting my mighty bed rail, the dog watched from his corner. At the front gate I called the chickens, who came running too. I shooed them out the front gate, and the dog was trapped in the buck pasture by himself.

I looked at him. He looked at me. He looked just like my next door neighbor's dog, except they keep their dog collared and on a rope when he's outside. This dog had no collar. I decided my best course of action was to call animal control. As I scanned the area to reassure myself that the dog was fenced in alone, he turned, hooked his front paws over the top of the fence, and hoisted himself up and over to the other side. This is a four-foot fence, and I'd never seen a dog do that. I watched him run around the trees, into the next door neighbor's yard, and onto their porch. It was mid-morning and no one was at home.

I went to check on the goats and count beaks. The boys (especially Billy) were upset but fine. Of the chickens I could only account for eleven instead of twelve. I went back to the buck pasture for another look. The dog had come back and was standing on the other side of the fence once again. He took off when I yelled. Concerned he might come back again, I walked the perimeter of the front pasture. My neighbor from across the street came over to tell me he had seen the whole thing, in case I needed a witness.

I found a couple scatterings of feathers on the ground, but couldn't find the missing chicken. I spent the rest of the day outside, keeping watch. The dog paced his own yard and ran around in the luxury of his freedom. Dan got home about the same time the dog's owners did, and he went to talk to them. Fortunately, they were apologetic rather than defensive. They had acted responsibly with this dog, and we understood that its getting loose was an accident. Still, accidents can result in dead livestock, because unless a dog is bred for working with livestock, it will (no matter how nice, well trained, or sweet) chase down small livestock to the death. It's not necessarily malicious, but instinctive, because dogs are, after all, predators by heredity.

Dan and I discussed the situation that evening, took a look at Craigslist, and made a decision. It was time to get a dog. My missing chicken eventually showed up at the coop, Billy remained skittish for several days, and the next time we saw the neighbor's dog, he sported a heavy new collar with a chain to replace the rope.

KRIS

The Craigslist ad said he was a seven-week-old livestock guardian mix. His father was a registered Bernese Mountain Dog, and his mama was half Great Pyrenees and half Anatolian Shepherd. All three of these breeds are classified as working dogs, the Pyrenees and Anatolians bred specifically as livestock guardian dogs (LGDs). Bernese Mountain Dogs are used extensively in Europe and some places in the U.S. as general purpose farm dogs.

The breed or breed mix was important to me. Genetic instincts seem to be key, and can rarely be trained out of an animal. Purebreds were out of our price range, so a mixed breed was something I considered carefully. It was also important that both of this puppy's parents had experience with livestock. Personality is a factor too, but this would be harder for me (in my LGD inexperience) to discern with a puppy. We needed a dog we could trust with our livestock, would live with the goats, and that had strong guarding instincts for the property in general.

One breed we'd looked at previously was the Australian Shepherd, also known as the Blue Heeler. The folks from whom we bought our

Pygmy bucks had one, and we really liked the size, personality, and look of the breed. When an ad came up for Blue Heeler puppies, we went to take a look. The gentleman was pretty straightforward with us, even asking what we wanted in a dog. He said Blue Heelers are primarily herding dogs, not guard dogs. If given the chance, especially with another dog around, they will start running the stock for the fun of it, often with disastrous results.

This new ad, then, was more hopeful. The gal with the livestock guardian mix puppies had about ten for sale, mostly female, with one of the two males already promised to the sire's owner. Dan wanted a male so I took the one that was left. After I paid her and carried him to the car she mentioned that he was the runt of the litter. Then she corrected herself, saying the runt had actually died, but that mine was the second smallest puppy in the batch. A small warning flag fluttered briefly in my mind as I loaded our new puppy into the dog carrier. Later I would wish I'd heeded it, but for now I was too excited over this little guy. I brought him home. Because he was born on Christmas Eve, we named him Kris.

There are two schools of thought regarding the training of livestock guardian dogs. Old school is for the puppy to have virtually no human contact for several weeks, in order to allow bonding with the stock. More current thinking is that the tendency to guard is more about genetic instinct. Some dogs naturally make better guardians than others, even within the LGD breeds themselves. This line of thinking says the puppy needs to be kept with the stock, but also needs human socialization. It needs to know basic commands, be leash trained, and be able to be transported for things like vet visits. Folks who raise LGDs now say the instinct to guard livestock just "kicks in." The newer thinking was definitely more appealing to us than the old. I made a comfy home for Kris in one of the two stalls in our goat shed. To become acquainted, Kris would have one stall, Jasmine and Surprise would have the other. The goats were not impressed.

Our first worrisome incident was when our new puppy came down with a respiratory infection known as "kennel cough." Fortunately we caught it before it developed into pneumonia, but it meant bringing him indoors while I treated him with antibiotics. He had to stay in the bathroom where I could keep the humidifier going. After that, he came in at night and on rainy days, something he did not like. I couldn't say that I blamed him and was glad he was an outdoor dog, but we felt he needed to use his energy for staying well, not for keeping his body temperature up in the cold and wet weather.

By the time he was eleven weeks old, he weighed twenty-two pounds and had become a gangly, all-legs puppy. He had taken to following the goats around and seemed to particularly like Surprise. She, on the other hand, had no use for him whatsoever, and presented him with the top of her head anytime he came up to her. Kris didn't know that in goat body language this means "mind your own beeswax," so he would try to lick her ears. Sadly, she had no appreciation for puppy kisses. She did seem to like his puppy food, however, and made a beeline for it when it was time for Kris to eat. Of course, this was something she was not allowed to have.

There were many lessons to be learned. These included learning not to get under foot, not to nip, and not to grab at hands, pants, skirts, and goat legs. I also wanted to teach him about barking if I could. Barking at things that don't belong here is okay, but barking at goats and chickens is not okay.

The hardest lesson was learning play versus not play. As a puppy, he was all about play. He didn't understand that there were seventeen other critters around which expected to be fed and tended to. On the other hand, neither did any of them. It would have been considerably easier on the humans if he liked to play fetch. Fetch is a wonderful way to exercise a dog and wear out a puppy. Oh, he'd play, but only if it was two, no more than three tosses of the ball or stick, and only if it didn't land more than six to ten feet away. Kris much preferred a game of tug or wrestle. This was okay during designated play times with the humans, but not okay when the goats were the object.

For the goats, it was initially "us against it," and they were slow to appreciate him. He wanted to play in ways they didn't. Gradually they began to tolerate him. Of the chickens, he decided to keep a respectful distance, for which I was glad. I definitely did not want him to think chickens were toys.

Our second worrisome incident occurred in early spring. One morning I found Kris with his hocks swollen. The only thing I could imagine was that it was from insect stings (we have ground bees). I was quite alarmed and immediately took him to the vet. The doctor agreed it might indeed be from insect bites or stings and put him on antihistamines. These had an almost immediate effect, and I was greatly relieved.

After that, everything seemed to go along pretty well. Kris was friendly and affectionate, smart but with a stubborn streak. We continued with our daily training sessions and gradually the goats stopped complaining when he tagged along. Then one day I found a message on my answering machine. Things were about to change.

KODY

The tale in which we became a two dog homestead started with a trip to the vet for the last of Kris's puppy shots. He weighed in at 43 pounds and was pronounced healthy and handsome. Except for getting into the car (he hated being put into the car) he actually behaved on the leash, for which I was very pleased. He could be stubborn at times and I was working hard to overcome that. After his vet visit I brought him home before going out to do some errands.

When I got home at lunchtime, I found a message on my answering machine. It was from Cathy, the receptionist at the vet's office. Shortly after I left, a gentleman had come in with a Great Pyrenees puppy for which he needed to find a home. He loved it dearly, but his wife wouldn't let him keep it. If he could just find a good home with plenty of room and love, he'd give it away. Cathy immediately thought of us.

We were in no way considering another dog, but at least had to think about it. I definitely would have said no if he had been anything else, but he was a full-blood Great Pyrenees. The gentleman who gave him to us was elderly and not without health problems. This was his

Kody and Kris were happy to be playmates.

dream dog and he was heartbroken to part with him. We told him he was welcome to come visit Kody any time he wanted.

Weighing in at 27.5 pounds, Dakota was three-and-a-half months old, one month younger than Kris. He had the happiest personality. The two puppies got along instantly, but not without adjustments. There were a few squabbles over food and toys because both were used to being an only puppy.

Kody didn't understand why he was here. He was used to being a house and couch pup, not a farm dog; outside time was run and play time. Because of that he was a little hyper the first couple of days, and there was a lot of barking and whining and running around any time he saw Dan or me. After a good romp with Kris he would wait by the gate to go "home."

His former owner did indeed come on occasion for a visit. Initially Kody was ready to leave with him. Gradually he would visit a little less, until it got to a quick greeting of "nice to see you, but I gotta go play!" He no longer minded sleeping in the barn and enjoyed his new life with us.

Are Two Dogs Better Than One?

Dogs are instinctively pack animals. They naturally like the company of other dogs. That makes it easy to keep more than one, and many folks like that their dogs have at least one canine companion. In our situation, however, Kris seemed to tolerate Kody more than truly befriend him.

By the time he was six months old, Kody was a very handsome fellow. He was happy-go-lucky, lovable, high-energy, playful, with a keen eye for anything amiss. He was quite the tease. If Kris had a toy, Kody loved to grab it just for fun. More often than not, Kris would get annoyed at this, and would growl and snap at Kody when his patience wore thin. Later I would realize that this was probably because he didn't feel all that well. Kody was always ready for a romp or a run, but Kris preferred to lie down.

Kody's only "fault" was that he loved to chase. Chasing Kris was okay with us, but chasing chickens, goats, and Riley the cat was not.

With two dogs, we noticed that Kris and Kody seemed to pay more attention to one another than to the humans.

Kody received many a scolding for this and spent quite a bit of "time out" in a stall. He was a smart dog and caught on quickly, but being a puppy, he had frequent slips in his self-control, especially if we weren't watching.

One thing I noticed was that Kris seemed to unlearn a lot of the lessons I'd tried to teach him. He knew better than to chase the goats, but when Kody playfully chased the young ones Kris couldn't help but join in. That pack instinct seemed too hard to resist. This greatly alarmed the goats, which I couldn't have.

The chickens were terrified of Kody. The sight of him coming would send them running in a squawking terror. This, of course, only excited Kody for a chase. When Mama Hen hatched her lone egg, the dogs were moved in with the bucks, just to keep Chicken Little safe. This gave the dogs access to the wooded buck browse, which they loved. The only problem was the occasional tick, but I assumed that the flea and tick medicine from the vet took care of that.

Poor Kris

One day we noticed Kris was limping. Upon taking a closer look, we discovered that his elbow was enlarged. He got to where he would lay around most of the day, not moving much and looking forlorn. The preliminary diagnosis was probable elbow dysplasia.

I say probable, because even though elbow and, especially, hip dysplasia are common in large breed dogs, and even though symptoms can be seen in puppies as young as four to ten months of age, a true diagnosis can't be made until the dog is twenty-four months old, according to the Orthopedic Foundation for Animals. X-rays are commonly used to diagnose, also CAT scans. Treatment is surgery and anti-inflammatory drugs. Arthritis always sets in, apparently even with surgery. It was rather devastating news for a puppy as young as Kris.

As I researched the problem I learned that heredity and diet are considered key factors (fat dogs are more prone to it). Besides the medical regimen, I wondered if there wasn't something else we could do. I presented the question to a holistic email goat group to which I belonged. A response came back quickly, about a Golden Retriever who had been diagnosed with dysplasia. Her owners started giving her a couple tablespoons of unflavored, unsweetened gelatin twice a day, simply mixing with her food. The results were reported to be amazing, and the dog continued to be active and pain-free seven years later.[1]

I started Kris on Knox brand unflavored gelatin on a Friday, one packet twice a day. By the following Tuesday he was running, jumping, and playing with Kody like he used to. The swelling in his elbow was gone and felt exactly like the other one. What a blessing! I found an economical bulk source for gelatin, where I could order five pounds at a time.

Why does gelatin work? Because it is rich in a simple, nonessential amino acid: glycine.[2] Traditional diets were richer in glycine than modern diets: head cheese, fried pig skins, pork ears and tails, chicken feet for soup broth, fish head soup, pork chops, chicken drumsticks, etc. All of these are rich in collagen, from which gelatin is made. Dogs would get glycine from eating skin, tendons, bones, and cartilage. Modern commercial dog foods contain little to none of these things, as is evidenced by the rise of arthritic problems in dogs. I learned about feeding our dogs raw, meaty bones, the emphasis being on raw, not cooked. Cooking causes the bones to become brittle with a potential for splintering into sharp fragments. This is why feeding bones to dogs (and cats) is frowned upon.

I started giving both dogs a raw meaty bone daily, plus added gelatin to their dog food. I had learned when Kris was a wee pup that he was allergic to the corn commonly used as a base for dry dog foods. At least I assumed it was an allergy because he would throw up when

When Kris wasn't feeling well, he didn't enjoy Kody's energy and enthusiasm.

fed most brands. Quick research on the internet told me that corn is a common allergen for dogs, so when I switched to a non-grain based dog kibble, the vomiting stopped. I would moisten their kibble with raw goat milk or raw beaten eggs and stir in the gelatin.

It made an amazing difference for Kris, although he had set-backs on occasion. While Kody grew at a slow, steady pace, Kris grew in spurts. We could almost see it and some mornings it seemed as though someone had swapped out our Kris during the night for a larger puppy! During the growth spurts he was mopey and limped. I would up his dose of gelatin, and he would usually be fine the next day – playful, happy and active.

This is obviously not an economical way to feed dogs, and, in fact, most of my grocery budget went for their foods. Still, it was more economical than the alternatives and I was grateful to have found a solution that did not require a lifetime of prescription medications and surgery with its less-than-hopeful prognosis. Plus, with our goal of feed self-sufficiency, I had reason to believe that eventually we would be able to support the dogs as well as ourselves from the homestead.

Good-Bye Kody

Chickens will be chickens just as dogs will be dogs. Over the years I learned quite a bit about barricading chickens out of places I didn't want them. Even so, I have never come up with a completely chicken-proof system. Although ours were afraid of the dogs, the braver (or stupider) ones found a way into the dogs' area despite my deterrents.

One Sunday afternoon Dan found Kody happily playing with a dead chicken. We didn't witness it, so I can't say for certain that he killed it, but it was highly likely considering his continued chicken chasing. We had a long talk about what to do. As a Great Pyrenees, Kody was a beautiful dog, but so far, he had not shown any tendency toward guarding. He was observant and watchful, but his fun-loving personality always won out. Like most intelligent dogs, he was headstrong. This is not a bad quality in a livestock guardian, who needs to be a thinker who will take action as needed. Livestock guardians also need to have some degree of self-control, which I didn't feel Kody was developing, if it was there to be developed. Kody's obedience was more of an easy-going compliance because he felt like it. Besides losing a chicken because we hadn't yet broken his chasing habit, I feared he would now associate the taste of chicken with the birds, a very bad situation indeed.

Some folks say that the way to break a chicken killer is to tie the dead chicken around the dog's neck. Left there until it rots and stinks, this is said to cure the dog of wanting to chase and catch chickens. I did not have what it took to do this. We could have kept Kody penned or tied up for the rest of his life, but I did not want to do that either. Kody had to go.

I was ridiculed by a reader of my blog for making this decision. I was told that getting a dog was a lifetime commitment, like it or not. I was told it was not right to blame the dog for its natural instincts and tendencies. The problem with this criticism was the assumptions it made. The situation had nothing to do with blame, but everything to do with being a responsible steward of all my animals, not just my dogs. My chickens and goats had just as much right to safety and well-being as the dogs did. Was it right to confine Kody because of his instincts? I certainly didn't think so. Neither could I be so arrogant as to assume that we were the only ones who could give him a good and loving home. I wanted what was best for all of our animals.

When Kody first came here we agreed that if we ever couldn't keep him, we would call his former owner. Dan did just that and within the hour both the gentleman and his wife were here to pick him up. In spite of originally not wanting Kody, the Mrs. had agreed without hesitation to take him back. Our sad situation had a happy ending for everyone except the chicken.

LOSING KRIS

Kris seemed to take Kody's absence in stride. I'm not sure if it was because he was glad to have all the human attention for himself once again, or because Kody tended to be more rambunctious than Kris liked. Fortunately he had more good days than bad, and we hoped that things would continue to improve.

His daily routine was to follow the goats around, but one day he either couldn't or wouldn't get up. It took both Dan and I to carry Kris's eighty pounds to the car to get him to the vet.

The official diagnosis was Lyme Disease. This is caught from ticks, which was unexpected, because I had been using flea and tick protection on both of the dogs. It had been a bad year for ticks and we'd found them on occasion on Kris, Kody, the goats, and us. I had decided to get some guinea fowl in the spring, hoping the anti-tick medication would be enough until then. Our vet also said Kris was anemic. This surprised me because of his diet, but I started adding organ meats as well as the raw bones.

Dogs have a 90% chance of full recovery if Lyme Disease is caught within one week. Kris rallied a few days after he started the antibiotics, but several days later he just seemed to give up; wouldn't move, wouldn't eat, and then he was gone. It was a real blow to both of us.

In a loss like this there is always some small comfort in the knowledge that the animal was given a good and loving home, and that as his steward we did our best. Even so, Kris always seemed to have something wrong. He never seemed fully healthy the entire time we had him. I was always treating him for something, and can't help but wonder if there wasn't more going on with him than we'll ever know.

Our One-Hour Dog

It took a long time before Dan was ready to get another dog. We talked about it on occasion, but I think it was seeing those coyotes passing through the yard that made the difference. After that, he gave me the go-ahead to look for another dog.

This time we decided to go with an older livestock guardian. These come up on Craigslist from time to time as needing re-homing. Folks who formerly had livestock would, for whatever reason, sell off their livestock and have a sad guardian with no livestock to guard. Like all animals, they are happiest when they are doing what they were created to do.

The reason not to get a puppy was two-fold. Most importantly was time. A puppy will take several years to mature enough to be able to deal with the coyotes in our area. The second was that I wasn't sure how to select a puppy that would grow into a good guardian. It wasn't that I minded all the work that went into training, but I'd learned from others that not all dogs of the guardian breeds actually make good guardians.

Our list of qualifications was short. Our new dog would need to be a guardian breed or breed mix, have experience working with livestock, and be comfortable around cats because we had three resident feline mousers. We knew the cats wouldn't much like having another dog around, but their opinion on the matter didn't count.

I kept an eye on Craigslist and would follow up on anything that looked promising. Some were too far away, some were too young, some were too expensive, some chased their owner's livestock. Finally, I found an ad for a two-and-a-half-year-old working Great Pyrenees. He had worked with cattle, pigs, chickens, goats, didn't mind cats, was known to go after coyotes, and was excellent with human children as well. He needed a new home because the youngest child in the family had developed allergies to him. He sounded absolutely perfect for us. They told me he needed good fences, because if he thought something was threatening his territory he'd jump the fence to go after it.

That Saturday afternoon, I drove 45 minutes to get him. What a sweetheart! I fell in love immediately. I brought him home, gave him a drink of water, and left him in my goat-showing pen next to the goats to get acquainted. I went inside to change clothes before taking him around to see the property and meet all our critters.

When I got back outside he was gone. I couldn't believe it! I frantically started looking around and finally saw him down in the woods at the back of the buck browse. He'd easily cleared two fences to get there. He did not respond to my calls; why should he? I was the lady who took him away from his home. I tried to follow but lost track of him.

I jumped in my car and drove the back roads looking for him, but he was nowhere to be seen. I came home and called the city police, county sheriff's office, and talked to animal control to give a description and my number in case someone found him. I also emailed his former owner to give her a heads up because I had no doubt he'd head for home.

I can't begin to explain how I felt about the whole thing: worried, foolish, uncertain, like beating my head against a wall. I desperately hoped he'd show up at his former owner's, but then what? Could he be taught to stay here? Would we have to surround the entire property with an eight foot fence? Should I try to give him back and see if they'd return my money?

I wish I had answers to those questions but I do not. Weeks and months went by but the dog still had not shown up at his former home. Nor did I ever hear from any of the agencies I contacted. Lost dog ads got no response. It's unsettling not to know what happened to him. We remain dogless to this day.

What I've Learned About LGDs

In our brief relationships with livestock guardian dogs, I have learned a lot. I cannot tell you how to choose a puppy with potential, nor give you tips on training them. In those areas I am still very much a novice. What I can tell you is that they are not like the dogs we had growing up. Those animals were pets and happy to be so. The LGD is not a pet, it's a working dog.

Pet owners are probably familiar with that joke about dogs and cats; the one where the dog says, "The humans care and provide for me, therefore they must be god." The cat, on the other hand, says, "The humans care and provide for me, therefore *I* must be god." We laugh because it says something true about the natures of these animals. Along those lines, I've heard folks say that they prefer dogs because cats are too independent. Livestock guardian breeds are not that type of dog.

Guardian breeds are intelligent, independent dogs, often described as headstrong and stubborn. They can bond with other animals such as goats, but are territorial as well. They seem to identify their territory as what needs to be guarded and protected. They need to learn basic skills, such as to stay, walk on a leash, and load into a vehicle. To avoid a battle of the wills (like I often had with Kris) the dog must have respect for the human. I cannot tell you how to successfully achieve that. I do believe that time, love, firmness, and consistency are key, but we never had a dog long enough to tell you how well this works.

The human must also respect and understand the dog. The goal is neither to cater to the dog so that it is unruly and uncontrollable, but neither is the goal forced obedience. Teamwork is what is wanted and this should be the goal. I've caught a glimpse of this with the dogs we've had.

Will we try again with a dog? What I can tell you is that if we do, I will definitely choose a puppy whose parents are working farm dogs, who has been around farm animals and has observed how its parents interact with the livestock. This would be true for either a livestock guardian dog or a general purpose watch dog.

For our particular homestead, a watch dog is probably more what we need than a guardian. Our property is small compared to a farm or ranch, and our animals are scattered in different paddocks. Watch dogs are not something I've researched, but I think I would go for a mid-sized, rather than large, dog. This is primarily for practical reasons. Dogs are expensive animals to keep when one considers vet visits, vaccinations, flea and tick preventatives, and especially good quality food. A huge dog needs a lot of food. While a good guard dog definitely earns its keep, I want to be able provide for its needs. That's part of being a good steward.

The other concern with larger breeds is that they are, in general, more prone to hip problems. Kris's problem was discouraging and, while not insurmountable, was not an experience I would like to repeat.

For the time being we remain without a dog, and I cannot say what the future will bring. One of these days the right dog will come along, and then our homestead will be more complete.

Guinea Tales

Keets

In July 2013, seventeen guinea fowl keets became the newest additions to the homestead. They arrived on an early Friday morning via the U.S. Postal Service. We decided to get them after having so many problems with ticks the previous year. Guinea fowl are supposed to be superior insect eaters; in fact, our neighbor claimed they would even eat fire ants. I figured I'd have to see that one to believe it.

Keets are smaller than standard chicken chicks, because guineas lay smaller eggs with two considered equivalent to one chicken egg. In my preparatory reading I had learned that they grow more quickly, however, and so need a higher protein starter feed. I couldn't find a specific guinea keet feed, so I got one for turkeys and quail.

They were interesting to watch in their new, but temporary, cardboard box home. They moved more quickly than chicks and darted around their box like little speed racers. I had quite a scare when I first saw them sleep. Baby chicks just squat down and nod off. Keets lay out flat with their legs sticking straight out. Shocking at first—"Are they still alive?"

By the end of the week they weren't all alive, and I was quite concerned that we lost six keets in so short a time. I found a few of them, just dead. They looked as though they'd been trampled, which seems to be a problem because keets tend to pile up. They run as a unit when they're uncertain about something and bunch up in a pile when they stop. I also had a few with persistent pasty bottom. Some resources say this is stress related, which makes sense considering they were shipped across country and had to adjust to new quarters. This problem only happened with the very littlest keets. No matter how diligent I was to keep them clean, none of these made it. That left us with eleven, all of which seemed to be healthy.

I found them to be fascinating birds. They would run like the dickens if I reached in to add food or a clean water bottle. I learned to announce myself by softly calling, "Guineaguineaguineas," so that they didn't panic when I approached their box. I just hoped the rest of my keets would do well.

Moving Day

It wasn't long before the keets outgrew their little brooder box in the laundry room. By the time they were two weeks old they were running sprint races around their box as well as running one another over. I figured it was time to give them a little more room. My plan was to give them a larger brooder area in what was to be their permanent home.

One of the challenges of keeping guinea fowl is getting them to roost indoors. These birds prefer to roost in trees or on top of their house rather than in it. There's a safety issue here, because it's possible to lose them to owls, foxes, raccoons, or opossums if not protected. Every guinea fowl how-to book and website I read stressed that in order for them to roost in their coop, they must be kept inside their new quarters for six weeks.

We carefully considered where we wanted to put our guinea fowl in order to start them right off in their permanent home. One option was to build them their own structure, but instead, we chose "Fort William," our log buck barn. It's farther away from the house than the chicken coop, but this is the area where we'd had trouble with ticks, so this is where they needed to be. I moved the bucks into the front

I started with a small area for the guineas and gradually enlarged it.

pasture for the summer, so the guineas had the little barn to themselves. I wasn't sure how well they'd like it when the goats moved back in, but I figured I'd cross that bridge when I got to it. I enclosed a corner of the little barn for them with cardboard. I would expand it as they grew until eventually they would have the entire barn and beyond.

A six-week confinement seemed like a long time. I wasn't sure about the guineas, but knew that chickens will peck and bloody one another when they're confined. I didn't want to take a chance that the guineas would do the same, so I bought them a cheap mirror, the kind you hang on the back of a door to get a full length view of yourself. I had read that guineas love to look at their own reflection, so I put it in their barn horizontally. They loved it and would spend hours admiring themselves. Another amusement I discovered was to hang a piece of baling twine from the ceiling to about their head height. This fascinated them and they spent a lot of time playing "grab-the-string." I hoped these two things would keep my guineas occupied while they awaited their big day.

To Tame or Not To Tame?

In my initial research I read quite a bit about how to tame guinea fowl.[1] Ours had an important job to do, so they were officially designated as working birds. I didn't plan to train or "tame" them so that they would let me pet them or sit in my lap, but I did need to get a head count at least once a day. I also needed to make sure they went into their shelter at night. I needed them trained rather than tamed, and I learned I could do that with white proso millet. This is the same kind of millet that is fed to parakeets, and apparently, guineas love it. That makes it an ideal reward for desired behavior.

I kept their millet, their "treat," in an old Parmesan cheese container. Our "training" consisted of me shaking the container and calling "Guineaguineaguinea." Then I would shake a little on the ground and step back. They absolutely loved it. My hope was that if they learned to come for their treat, I could sprinkle it inside the buck barn in the evening to lure them inside for the night.

This actually worked quite well. Between routine handling and the millet, I felt like quite the guinea wrangler.

TEMPTATION

Six weeks inside, no going out. That's what I read. By the time the keets were almost seven weeks old they had the entire little buck barn for their territory. I would check on them several times a day and find them congregating by the door, looking longingly at the great outdoors as though they knew they had a mission out there. I was so tempted to let them out, yet I heeded the advice of those who have gone before. Those who succumbed to that temptation all seemed to regret it later, because they would have problems with their guineas roosting outdoors. As badly as I wanted my guineas to be happy, the potential consequences of letting them out prematurely kept me in check. Having lost one chicken which insisted on roosting at night in a cedar tree rather than the chicken coop, I was motivated to see this through.

I had two more weeks to keep them indoors. Then they'd be eight weeks old, having spent the recommended full six weeks in their permanent home. I just hoped it would be enough.

The Big Out

Finally the big day arrived, the day the guineas were to be allowed outside. I decided to start with just a few hours in the afternoon. If I was expecting them to pour excitedly out of the barn into the great outdoors, I was wrong! It took nearly three hours before they finally dared to cross the barn door threshold. Then, when it was time to go back in, they were uncertain about that and had to be coaxed. This is where our training sessions paid off. The sound of the millet being shaken in the container immediately got their attention and they were ready to follow me anywhere!

The next day I let them out around noon. It wasn't long before they were out and about, inspecting and exploring. They were the funniest things; cautious and always moving as a unit like a school of fish. They made their way to the front of their quarter-acre pasture and discovered goats and chickens! Best of all, they made their way back to the barn all by themselves in the early evening. Food, water, and their beloved mirror were in the barn and, of course, that was where they got their "treat."

Introductions of new species are always interesting on the homestead.

On the following day I let them out a little earlier, and they spent the day in busy search-and-eat activity. I read that guinea fowl have quite a territorial range, and I planned to gradually expand where they were allowed to roam. That assumed they didn't figure out how to jump fences before that. As long as they could find their way home at night, I'd be happy.

The Rules

Three weeks later my guineas were still going inside at night to roost. I was extremely proud of myself for that, considering the species is so notorious for not cooperating. In return for this compliance, however, I was expected to follow a couple of rules.

Rule #1: Be on time. Like all other animals, guineas like routine. Ours went like this: my chores began at first light when I let the chickens out of their coop first, then I let out the guineas. I'd check water and feeder levels, toss some scratch into the chicken yard, and give the guineas their millet. They'd clean that up, pour out the door, and then fly up to the top of the buck barn roof. After their morning conference, they'd set up a squawk and fly off the roof and into the corn patch. They looked like ducks coming in for a pond landing.

After evening milking and before dusk, I'd make my final check on the guineas and put them up for the night. They were already in the barn, but when they heard the gate, a few heads would poke out the door to make sure it was me, and then they'd run excitedly back inside.

They had been visiting their mirror while waiting but were eager for their evening treat. Once again I checked water and feeder levels, and gave them a generous sprinkling of millet before closing them in for the night.

One evening, Dan took me out to dinner. It wasn't a late night out, but we didn't get home until after dark. I was a little concerned about the guineas, so I immediately went to check on them. I arrived by flashlight at their barn and was alarmed to discover it was empty. I called in my customary "Guineaguineaguinea," and heard a stampede thundering from one side of the metal roof to the other. I went out, shined the flashlight up to the roof, and there they were—ten guineas all peering down at me. I could not coax them down for anything. It didn't matter how much treat I sprinkled on the ground, they were staying put. Eventually I had no choice but to leave them there.

I was relieved to see they were still there the next morning. They flew down when I arrived, gobbled up their treat, and went on about their guinea business as usual. Happily there were still ten, but I was worried this would set up a new pattern, that of sleeping on the roof instead of coming inside to roost.

That night I arrived "on time" and they were waiting for me inside. That was the case every night unless I arrived too early. Then they'd all run out again as if they weren't ready to roost. As long as I arrived just before dusk, they were ready to bed down and I knew they were safe for the night. Lesson learned.

Rule #2: Dress appropriately. "Appropriately" to an animal does not mean the same thing that it does to a human. To a human it means to dress according to the situation. To an animal it means to dress as expected. Part of the routine is that you, the keeper, must show up as they expect to see you.

Usually I do all my chores in work clothes: an old t-shirt with optional old jacket and an old skirt or old jeans. If, for example, it's raining hard and I wear a poncho to do chores, I can expect some nervousness amongst our animals until I call out to reassure them. Once they recognize my voice, they calm down. Not so with the guineas. One rainy evening I arrived in a red poncho which sent an immediate alarm through the guinea ranks. I tried to reassure them that it was me, but they were in such a frantic panic, they couldn't hear my voice. I thought I was going to have ten guinea heart attacks and ten dead guineas! Lesson learned.

I suppose it might be argued that the guineas aren't the ones who are trained, rather, I am. Actually I have no problem with that. As long as they're safe and healthy, I'm willing to do whatever it takes no matter who's in charge.

THE TRUTH ABOUT GUINEA FOWL

I don't think anyone will ever accuse guinea fowl of being beautiful birds. What they are very good at, is making their presence known. I can see why folks say they are good "watch dogs." The real scoop is that they carry on about everything, including, but not limited to, anything unusual in the yard, or the neighbor's yard, or the neighbor's neighbor's yard.

The most common question folks ask about guineas is, "Are they really as noisy as they are reputed to be?" The answer is yes. They are loud and noisy sometimes, sounding like rusty hinges on an old screen door. At other times, they chirp and twitter amongst themselves as melodiously as songbirds. The other problem is that they roam.

The noise and roaming were always potential problems because we have several neighbors right across the street. As long as the guineas stayed away from the road, no problem. True, they'd often hop the fences to check things out next door, but there were never complaints. The neighbors on the one side have lots of acreage and likely never noticed. Our neighbors on the other side have their own chickens, so occasional stray birds on either side of the fence (theirs or ours) are mutually accepted.

When the guineas wandered too far we'd fetch them back. They would come for chicken scratch or millet when called, and could usually be distracted by being allowed into the chicken yard. For some reason I couldn't fathom, they loved the chicken yard. They would happily spend the afternoon there (much to the indignation of the chickens). Even so, it was a chore to keep retrieving them.

When it was time to eat, they'd let me know! One thing I learned was that they loved chicken scratch as much as they loved their millet. Any time I called the chickens, the guineas would come running too.

When they were ready to go back to their barn in the evening, they would fuss at me to get going. As soon as I grabbed the feed can and a scoop of scratch, it was a guinea race back home. As long as I stuck to the routine and the rules, almost everything was fine.

What wasn't fine had nothing to do with me and everything to do with guinea social structure. There were two of them who appeared not to have made even the bottom ranks of the guinea pecking order. They seemed to be guinea rejects.

These two were continually chased away by the others and not even allowed into the guinea coop at night. It began to take forever to get the guineas put up in the evening, because the last one was afraid to go in with the others. I'd actually seen them attack it several times. Eventually it would go in, and while I felt badly for it, I thought this was safer than leaving it out all night as easy pickings for some predator.

THE BEGINNING OF THE END

One day the guineas were being particularly naughty. They decided they were going to go across the street. We'd already had several go-rounds with one of our neighbors about their dog doing its business on our property and wanted to keep things amicable. Having our guinea flock continually exploring others' neat, trim, suburban-looking yards didn't seem a good gesture toward neighborliness.

That day, I had to go get the guineas several times. None of the usual distractions worked, and the last time I went to retrieve them, they balked when they got into the street. To make matters worse, a pick-up truck came barreling over the hill and had to stop because of them. I'm out there waving my arms and trying to herd them to our side of the road, they're all screaming at the truck, and the driver is making faces and hand gestures to let me know that having to wait 20 seconds for me to get my birds out of the road is not acceptable.

As I herded them toward the back of our property I debated what to do. As much as we loved and wanted the guineas, their noisy wandering was getting to be too much. Keeping track of them was becoming a full-time job. I began to understand why the job of herding animals was often given to youngsters in days gone by. Still, I needed to do something immediately.

I took them back to their little barn and used chicken scratch to lure them inside. I closed the door, posted an ad for free guineas on Craigslist, and waited by the phone. Eight of them were locked up with two still on the loose. I figured I'd keep those two which would be a more manageable number for us than ten.

The folks who took them were trying to get their own homestead started. They didn't live close so there was no chance of the guineas trying to return home. It was hard for me to watch them go and I kept hoping a few others would get away so they could stay too. With only two birds left, it was amazing how quiet the place was.

WHAT HAPPENED IN THE END

At evening chore time a couple of weeks later, I went to feed our remaining two guineas but they were gone. Dan had seen them at about 2:30 that afternoon and I discovered they were missing around 5 p.m., so they disappeared within a several hour timespan. We'd been outdoors and never heard them holler and fuss, so we had no clue as to what happened. Did something get them? Did they decide to leave? Did a fed up neighbor pick them off? Did a hunter think they were wild turkeys or oddball pheasants?

A few nights after they disappeared Dan awoke in the middle of the night and went to look out the window. He thought he saw something run by and went outside to investigate. While he stood there in the dark, something came running across the road. At first he thought it might be a large fox. A second one followed and he shined the flashlight on them. They were coyotes, likely the same ones our neighbor told us about the previous summer. I don't know much about the hunting habits of coyotes, but that seemed the most probable reason for the last two guineas' disappearance.

Of all our critters, I think the guineas have been my favorite. Even though they were ugly, noisy, and wouldn't stay put, they somehow won a place in my heart. Because they refused to stay put, they were probably the most frustrating critters we've had so far, but they were also the most amusing and entertaining. We got them because of their reputation to eat ticks. What we discovered is that they are unique and remarkable birds. They have personality, and opinions!

Did they do the job for which we "hired" them? I'd have to say that since the guineas lived here, we've found fewer ticks on our other critters and ourselves. Perhaps if we'd had them longer they would have eliminated our tick problem, but that's a guess on my part.

Would I consider getting guinea fowl again? I think about that. Maybe someday when I have a broody hen, I'll see if I can find someone around here with keets for sale and graft a few on to her. I'll just have to wait and see.

Kitty Tales

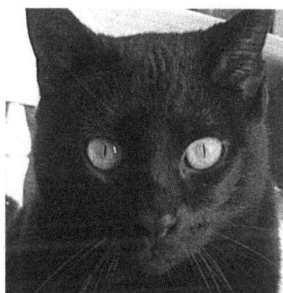

THE ONE WHO CAME BEFORE

That we even have cats is something of a tale to tell, because when I first met Dan he did not like cats. He was a self-proclaimed "cat-hater" and proud of it. As many a former cat-hater will tell you, however, it's just a matter of knowing the right cat. This is the tale of how Dan met the right cat.

Long before we moved to our homestead, we lived in what I call "rural suburbia." The area was former farmland which had been divided and parceled out first to the farmers' grown-up children, later to land developers. As factories moved in, fewer folks farmed and more took up an urban style of life. The area certainly looked rural enough, but the homes and yards were more suburban in character. We rented a 30-year-old ranch house on an acre or so of land, kept a large garden, but no critters. Cats were not initially welcome in the yard, until it occurred to us that they help control the mouse, mole, and snake populations. Pretty soon we became familiar with all of the neighborhood cats.

There was a pretty little black and white lady cat with tuxedo markings. Her name was Susan. She lived two doors down the hill and made dainty trips through the garden. There was a big white tom cat with a ginger tabby tail and a large ginger patch over his left shoulder. We called him Whitey Tiger Tail. Whitey Tiger Tail was big, fast, and chased all the other cats away. He chased squirrels too and sometimes caught them. He lived up the hill across the street. Then, there was an old battle-scarred gray tabby tom who stomped across the yard like a bull dozer. We called him Tough Guy. We later found out that he lived at Susan's house and that his humans called him Clarence. We liked "Tough Guy" better.

One day we saw a new cat. He was a young gray tabby who began to make trips through the yard from time to time. He looked about 6 months old. He was skinny and afraid of people and the other cats. He looked so much like Tough Guy that we started to call him Junior.

Dan commented that he thought it wasn't fair that the other cats always chased Junior away. He said that maybe if Junior got bigger and put on some weight, then he could take care of himself better. The next time we saw Junior in the garden, I took him some leftover chicken. Junior was scared and wanted to run away until he smelled the chicken in my hand. I set it down on the ground and stepped back. The poor little guy was very hungry and he eagerly went to eat the chicken. As I stood there watching, Tough Guy came by on his way up the hill. He could smell the chicken too and came over to see what was going on. He really wanted to take that chicken away from Junior, but I wouldn't let him. Tough Guy and I squatted there side by side, watching Junior eat.

Junior must have known he was being protected from Tough Guy, because after he finished eating he allowed me to present my hand for him to sniff. Tough Guy sauntered away, but Junior stayed there and took a bath. After that I put out a bowl of cat food for Junior every day. Junior would come and eat it, but the other cats would come and eat it too, especially Whitey Tiger Tail, who would chase Junior away.

Gradually I moved the bowl closer and closer to the house. Junior was nervous about this at first, but soon began to feel safe close to the house, where we wouldn't let Whitey Tiger Tail or Tough Guy eat his food if we could help it. I eventually moved his bowl into the carport next to the back door. Junior knew he always had something to eat. Whenever he wanted a drink, he helped himself to the bird bath.

As winter approached and the weather began to get colder, Dan was concerned that Junior would be too cold at night. He didn't want a cat in the house, so he fixed up a cardboard box for Junior to sleep in. He chose a cozy-sized box with high sides to protect Junior from the cold wind. Then he put a scrap of carpeting in it and some old towels to keep Junior warm. He put it in a secure spot on top of a pile of boxes in the carport, so Junior could climb up high and feel safe.

Junior loved his bed. He would sleep in it at night and sit in it and look around during the day, when he wasn't off hunting and exploring. Sometimes he would just sit there and watch the other cats passing through the carport. They never seemed to notice that he was there.

One day Whitey Tiger Tail discovered Junior's box. He jumped up into it, sniffed it all over, and then sprayed it. After that Junior never got into his box again.

Whitey Tiger Tail always seemed to be on the lookout for Junior and would chase him away whenever he saw him. But Junior knew that if any of the family was around, they would protect him from Whitey Tiger Tail and he would be safe.

One day I heard a desperate, wailing yowl from right outside the back door. I ran to the door and threw it open. There at eye level, was Junior, who had climbed up the screen door and was hanging on for dear life. I saw Whitey Tiger Tail too, slinking away under the next door neighbor's bushes. After that, Dan decided that Junior would be safer if he could come inside. So we began to invite him in. At first, Junior wasn't too sure about being in the house. He was polite, but he kept an eye on the back door. He was very happy to find his bowl of cat food beside the door, but he was too nervous to eat. We talked softly to him and he finally settled down to eat some food.

After a few minutes our Audubon bird clock struck noon. The 12 o'clock bird was a Great Horned Owl. It hooted the noon hour loudly and Junior panicked. I opened the door and let the frantic little guy out. That was his first visit.

Gradually Junior became a regular part of the household. At first he would only stay for short visits and wouldn't venture out of the kitchen. But as he got used to the hoots and chirps of the bird clock, and got used to being in the house, he would stay longer and longer.

One thing that puzzled Junior was why we kept rolling little balls at him and dangling yarn in front of his face. We were puzzled too, because he wouldn't bat at or chase the toys we offered to him. We began to think that Junior didn't know how to play; that because he had been on his own all his life, he had been too busy trying to survive to have learned how to play. So Dan began to spend time enticing Junior with yarn and catnip mice. Once Junior caught on, they had great fun playing ambush, pounce, and chase.

In the kitchen I had a string of jingle bells hanging on the inside door knob of the back door. Whenever Junior wanted out, he would reach up and bat the bells. When we heard the bells, we knew to come and let him out. The more comfortable he became in the house, the less he wanted to go out. However, we didn't want to leave Junior

Junior gradually went from feral cat to house cat.

inside when no one was at home. He wasn't used to being in the house for long periods of time and we were afraid he would panic if he couldn't get out. Anytime we were ready to leave, we would jingle the bells on the door knob and call, "Kitty, kitty."

At first, Junior would come to be let out when he heard us call and jingle the bells. Gradually though, he became less and less interested in going out when we wanted him to. He began to ignore us, especially when he was busy taking a nap, so I started to bribe him with chicken.

Junior loved chicken. He knew where the chicken was kept too, in the big white box full of food where it is always winter. Every time someone opened the door to the big white box he would come running and ask for chicken. If he didn't get it, he would try to jump inside to look for chicken himself.

One day there was a terrible storm. There was no thunder or lightning, but the sky was murky, the wind was howling, and the rain was pouring down. Junior didn't know it was storming, so he batted the jingle bells to go out. When I came to the door and looked outside, I told him that he might want to wait until later. He rubbed my legs, gave me his sweetest kitty face, and cranked up his purr to high volume. So I opened the door. Ordinarily, Junior didn't mind the rain. He had spent most of his life outdoors and could even sleep in the rain. This day, however, Junior took one look out the door at the blowing wall of rain and stopped short. His purring stopped abruptly. He stood in the doorway and stared outside. Then he casually turned around, walked back through the kitchen, jumped up onto the desk, and settled down to take a nap.

As Junior, the timid feral cat, changed, Dan, the confirmed cat-hater, changed as well. As Junior discovered that he liked having a human family, Dan discovered that he liked having a cat. We decided it was time to take Junior to the vet for a check-up and vaccinations. I worried that he might panic in the car, so we decided that as soon as we could get a cat carrier, I would make an appointment. Then something happened that changed everything.

Junior always spent the night outside and every morning he would be waiting by the back door to come in for his breakfast. One morning he wasn't there. I didn't worry about it because Junior was often gone for long periods of time. Sometimes he would not come home for several days, though these absences were becoming less frequent.

The next morning I opened the door expecting to see Junior, but he still wasn't there. I called, but no Junior. Dan had to go to work, so I set out to look for him.

Our house was about ten miles from a small town in the foothills of the Blue Ridge Mountains. We lived on a long curving road and had neighbors on both sides and across the street. Behind the back yard was a thickly wooded hill which sloped down to a small creek. We had seen opossums and deer in the woods, an occasional dog, and of course, cats. Beyond the woods and across the creek was a dirt road with more homes on it. In the winter you could see some of the homes through the bare trees, but in the summertime all you could see was the woods. There was a path through the woods which ran behind the backyards of the homes on our street. I walked through the woods along this path calling for Junior, trying to think of where he might be. When I returned home without him, I started to feel worried.

Every day we looked for Junior. Either Dan or I went outside frequently to call him. We looked for him everywhere we went, everywhere we drove. We looked on the sides of the road in fear he might have been hit by a car, but no Junior. Dan called on all the neighbors and asked if anyone had seen him, but no one had. I took a trip to the county pound in hopes that someone had found him and taken him there, but no Junior.

The days turned into weeks and the weeks into months, but Junior never came home. Of course we speculated for a long time about what had happened to him. The suspicious thing was that, not only did Junior disappear, but all the neighborhood cats disappeared. All except Susan, who never wandered very far from her home two doors away.

ODE TO JUNIOR

Where does Junior go at night?
I see him all the day.
But when the moon is high and bright,
His head, where does it lay?

Is there some secret hiding place,
That no one knows but he?
A field or open prairie space,
Or perhaps a forest tree.

Where does Junior go at night?
I sure would like to know.
My curiosity is ripe.
But how could I follow?

His size, his frame is oh, so small,
He's not an ape you see.
A kitten, tiger striped, that's all,
Without a pedigree.

Where does Junior go at night?
The future it might say.
For now, I'll settle at the sight,
Of Junior in the day.

By Daniel Tate

Rascal

One day I mentioned getting another cat. Dan said he'd had it with cats because losing Junior had been too painful. No cat could take Junior's place. Eventually, I suggested getting a rabbit. Rabbits can be litter-box-trained and make good house pets, plus, if I got an Angora rabbit I could have Angora fiber to spin. Dan thought about it and finally agreed.

Rudy was a white, ruby-eyed (or albino) French Angora rabbit. He was six weeks old, very friendly, very curious, liked to be petted, and liked to sit on my lap. He had his own hutch in the family room in the basement, but he was allowed to run around when someone was home to keep an eye on him.

We all liked Rudy, and Dan even tried to play with him the same way he had played with Junior. But somehow it wasn't the same. One day he said, "Rabbits don't play like cats, do they?" I could only agree. We didn't say anything else for awhile. Finally Dan said, "We need another cat."

"Well," I said, trying not to let my excitement show, "I know someone whose cat had kittens not too long ago. I'll ask her about them if you like." Dan agreed, so I gave her a call.

When my friend's kittens were old enough, she put all six of them into a big cage, put the cage in the back of her Chevy Suburban, and drove over so that I could pick one. Dan had to be at work that day, but he gave very specific instructions for the choosing of our cat. He wanted a male, and he wanted him to have lots of personality.

The kittens were beautiful. There were four girls and two boys. I thought the girls were the prettiest, because they had beautiful markings and coloring. I was very tempted to choose one of them, but since Dan was a recently reformed cat-hater, I decided it might be a better idea to pick the one closest to what he asked for. That left the two boys: a little black kitten with a long tail, and his brother, a black and white. Both were very feisty, but friendly, and it was difficult to choose between the two. I watched them play for a long time and noticed that the black and white liked to hiss. He would hiss whenever the others pounced on him. He would hiss when they chewed on his feet, tail, or ears. I didn't think all that hissing would go over very well with Dan, so I chose the little black kitten.

When Dan got home from work, the first thing he wanted to do was see the new kitten. He seemed surprised. "He's black!" "Not all cats are tiger striped like Junior," I said and we laughed.

We showed our new kitten his new food dish, water bowl, and litter box. After he ate some of his crunchy kitten kibbles and had a drink, he set out to explore. Dan got out some string and wiggled it across the floor. They played pounce and chase for the longest time. We decided to name him Rascal.

As the game began to wind down, however, little Rascal began to cry. This puzzled Dan and me because everything seemed to be going exceedingly well. Our bemusement turned to concern as Rascal began to run around and yowl. Something was clearly wrong, but what in the world was it?

Dan and I looked at one another, not knowing what to make of it. Rascal dashed under the little table next to the sofa and pooped. We gasped, and then there was silence while we stood there staring at one much relieved little cat and his little pile. No one said a word. The longer Dan stared, the more concerned I became, thinking this might very well be the end of Dan's new career as a reformed cat-hater.

Suddenly, Dan laughed. "He couldn't remember where his litter box was!" he said. I laughed too, but it was more in relief than amusement. Dan showed Rascal his litter box once again, I cleaned up the mess, and that was Rascal's first day as our family cat.

Rascal turned out to be a good hunter, and I'd seen him sit motionless in the garden for hours on end. His patience always seemed to be rewarded, because I would find mouse and shrew remains scattered about the garden. He was probably the most intelligent cat we'd ever had: he knew that the word "vegetables" meant something inedible for cats, and that his favorite stuffed toy was The Monkey. When asked, "Where's The Monkey?" he could find it wherever it was hidden and bring it to us. His greatest accomplishment, however, was that he could count to three. At least we thought he could count to three, because when he was told he would get three kitty treats he would walk away after he got the third one.

Rascal considered his social status in the family to be below Dan's, equal to mine and my daughter's, but above my son's. Any time Nolan had The Monkey, Rascal would take it away from him and hide it. One time he claimed the top level of Nolan's bunk bed as his personal sleeping space and defended it if Nolan tried to sleep there himself. He also assigned Nolan as his personal doorman at 6 a.m., and would howl outside Nolan's bedroom window at that time to be let in. In all things, Rascal had to get in the last word (meow), and had a bit of a temper too. If he lost in any cat games he would chase me around the house and nip at my ankles. It didn't matter if I had been participating in the game or not!

One of Rascal's games was one he played with Nolan. It was called "Rescue Me." Rascal would climb onto the roof of the house by means of the columnar arborvitae next to the carport. From there he would yowl until Nolan came out and got the ladder. Rascal would allow himself to be rescued and carried back down to the ground. He played this game frequently, whether Nolan was in the mood or not.

One day Nolan was off at a 4-H event when Rascal decided to play this game. I was in the kitchen, but when I heard him yowling I knew what was up. Not being in the mood to play myself I went outside, looked up at him peering over the edge of the roof in anticipation of Nolan, and told him, "Nolan's not home. You're going to have to get down yourself." I went back into the house and within a few minutes Rascal was at the back door, meowing to be let in. That's when we figured out it was a game, which I believe disgruntled Nolan greatly.

Rascal was well-traveled, having made two long-distance moves with us. Landing in a small second story apartment after the second move was a big change for all of us. We were used to open, outdoor spaces, not the feeling of being cooped up. Because of the parking lots and traffic, we decided he would not be allowed outdoors. He became very depressed, so I decided that the thing to do was to get another cat.

LITTLE CAT ZEE

It was a hot August day when I went to the animal shelter to look for another cat. Some of the cats had reduced adoption fees listed on their cages. Regular fees were $79, but cats which had been shelter residents for a long time were $25. This included current vaccinations plus spaying or neutering. This not only makes acquiring a pet more affordable, but also gets them out of a cage and into a home. I loved the idea of being able to liberate some poor kitty from a cage.

I knew what I was looking for. I knew Rascal would not be pleased, so I needed a cat that wouldn't be intimidated by him. There was a young tortoiseshell who caught my eye, because she would try to swat at me every time I walked by. She had no reduced fees listed on her cage, however, so I moved on. Still, I kept coming back to her, and she would bat at my shirt or purse strap every time I stopped at her cage. Money was tight, however, and I felt obligated to get the best bargain I could, so after each stop I would keep looking. At one point a family with a little girl asked about the little tortoiseshell. I overheard the volunteer telling them the adoption fees were $79. She must not be on sale, I thought. The family moved on and chose another cat.

*Catzee displayed what tortoiseshell cat owners refer to as "tortitude."
This refers to a fearless boldness which is common to this color type.*

I had narrowed my choices down to a couple of prospects, but finally admitted to myself that the little tortie was The One. That being the case, I would go ahead and pay full fees, because the right cat is the right cat. Imagine my surprise when the same volunteer told me this little cat was $25! I didn't question it. I completed the adoption and loaded my new kitty in the carrier I'd brought. It didn't take long at home, however, to realize that the adoption fees were a matter of the right price for the right person. This little cat was a spitfire and would have had that little girl scratched to shreds in no time. This little cat needed an adult caregiver.

I wanted to name our new cat Trudy, but Dan nixed that and named her Little Cat Zee. I called her Catzee (rhymes with Patsy) for short. She wasn't exactly wild, but she was so full of energy that she'd play chase until she was gasping for breath. Even then she wouldn't stop; I would have to make her stop. She was a natural hunter and better than a fly swatter. Our apartment remained fly- and bug-free from that time onward.

Our second story apartment had a balcony. To let Rascal go out, we had installed a sliding door insert with a built-in cat door. He would often make his retreat from his energetic "sister" onto the balcony. The thing Catzee wanted more than anything in the world was to go outside. It took her awhile to figure out the cat door, but she soon did and was on the balcony right along with him. Shortly after this, she discovered the flag pole attached to the balcony rail. It was attached at a right angle to the balcony, put there to display one of those decorative garden flags. It also made something similar to a "tight rope" for a cat to walk out on. And walk out she did. From there it was a quick flying leap onto the bushes below and she was free!

I witnessed the first leap and made a frantic run out the front door and down the stairs. I could see no sign of her. The kids playing in the parking lot watched me for a few minutes and then asked if I was looking for the camouflage cat. Yes! They pointed out the direction in which she'd headed and I soon brought her back inside.

That incident began a series of escape attempts which resulted in Catzee being banned from the balcony. Her next attempts were mad dashes through the front door whenever it was opened. Usually we caught her in time, but one escape seemed to cure her of this trick. She had dashed out the door and successfully evaded capture. We were expecting her to make a beeline for the stairs but instead, she ran between two balusters and jumped! There were no soft bushes to catch her there, and she landed on the concrete below. Fortunately she wasn't injured, but she never tried that trick again.

After that I bought a harness and leash and took her on daily walks outside. After Catzee had a turn I would take Rascal out for a walk too. In many ways the three of us were in the same boat, all feeling desperately trapped in that apartment. Fortunately the move to our new homestead was imminent, and we could look forward to freedom at last.

DISAPPEARED

As badly as both cats wanted to go outside, we were slow to let them after we got settled on our new acreage. When we did, it was initially on harness and leash until they became familiar with our yard and learned their way around. Rascal was a seasoned outdoor cat, with street smarts, so to speak, but as far as we knew, Catzee knew nothing about outside except that getting there was the ultimate goal in life.

When she finally was allowed out on her own she proved to be an excellent mouser, with a record of three catches in one day. She would spend the day hunting. When she found me working outside she was always happy to see me, full of purrs and requests to be petted. She always came in early and went to bed around 8 p.m., preferring day hunting. This routine was the pattern of Catzee's days.

One morning in late August I let her out as usual. I didn't see her once all that day, which was odd. When she didn't show up that evening, we were worried. We called and searched, but no Catzee. The rest of the week was spent doing all the things one does to find a missing pet: signs, posters, phone calls, and trips to the area animal shelters. The not knowing was the hardest part and, for Catzee, there were neither clues nor leads. Occasionally one hears miraculous stories of animals that have turned up after being gone for months. Because I was hoping for one of those miracles it took a long time for me to stop expecting to see her waiting on the back porch. Almost a year later I would continue to call for her on occasion, that's how much I missed that cat.

Cat lovers may be quick to want to say, "The only safe cat is an indoor cat." It is true that an indoor cat is protected from cars and traffic, larger predators, getting lost, and catnapping. It does not protect them from all accidents nor disease. It is also true that by their natures, cats are hunters. They make a purposeful contribution on the farm or homestead by helping keep rodents under control. This in turn helps keep down the rodent-hunting snake population.

All that aside, I never saw Catzee happier than she was outside, on the prowl for her next catch. We might still have her if I'd kept her inside, but was sacrificing her happiness worth that? Who among us wants to be imprisoned for the sake of "safety?" Every critter on our homestead has a purpose, and I cannot see denying anything its designed purpose.

The only one who didn't seem to miss her was Rascal. He seemed to look for her on occasion, but he quickly resumed his role as only cat. Unfortunately for him, this role wasn't to last for long.

NOT IMPRESSED WITH CHICKENS

Rascal was not impressed with my chickens. The only curiosity he showed was when they were newly acquired chicks. It was February and I had the brooder box indoors under a heat lamp. What cat could resist the ongoing "Peep, peep!" coming from that box? At his first peek I told him "No!," firmly repeating it several times for emphasis. After that, he ignored the box altogether. Even so, I kept the door to that room closed when I wasn't in there.

He was not inexperienced with other species. Rascal had seemingly developed a friendship with Rudy, my Angora rabbit. Both being youngsters, they would chase each other around the basement and then flop down together for a rest.

Rascal had a sense of duty when it came to rabbit chores. I would tell him, "Time to feed the bunnies!" and he would always accompany me to the hutches, although he would try to distract me from tending to the rabbits. His favorite trick was to climb the dogwood tree which shaded the hutches, climb out onto a wobbly limb, and then meow pathetically as if in need of rescue. If I was close enough, he would grab at me. The only exception to all of this was when it was raining or snowing. Rascal would dutifully come outdoors with me at rabbit chore time, but would sit in the carport and wait until I was done. As soon as he saw me coming back round the corner, he would immediately head for the back door and ask to be let in.

Rascal seemed to have a similar opinion of the chickens, that is, that they were his competition for my attention and that he was going to win. To his credit, he never stalked, or chased a chicken. Instead, he patrolled the yard for mice, moles, shrews, and chipmunks. He understood that, "Time to feed the chickens!" was his cue to follow me to the chicken yard, where he would quickly lose interest and go about his kitty business.

The chickens were always cautious of him, but, eventually, each accepted the other as part of the homestead landscape. All cats since have been the same, except Sam. He learned that by making a loping pass at the chickens he could send them into a tizzy, which he seemed to think was great fun. Some of my goat kids have caught on to this trick as well, making mock charges at the hens until they take off in a squawking huff. It all makes for interesting observations of the dynamics amongst the species on the homestead.

RILEY AND KATY

Rascal was diagnosed with feline lymphoma while we were still homestead-hunting apartment dwellers. Thanks to surgery and chemotherapy he outlived his prognosis by several years, although he was not the cat he used to be. He had become slower and a lot more cautious. Being able to go outdoors again really seemed to help, but he had his troubles. I once watched him try to race up a tree the way he did when he was young. When he lost his balance and fell it broke my heart.

Now he had out-survived Catzee, who was supposed to carry on mousing duties after he was gone. I figured it would be easier on us to have another cat already established on the homestead, than to try to replace Rascal under the weight of the sadness of losing him. After Catzee disappeared, it took a while to consider her replacement.

This time we found free kittens in the want ads. Considering how grumpy Rascal had become because of his health issues, I convinced Dan that we would be better off getting two kittens rather than one. A kitten is going to want to romp, and play, and ambush, and pounce; better with another kitten than with Rascal. There was always the possibility they would gang up on him, but we knew Rascal would quickly and decisively put them in their place.

Our new kittens were a silvery gray tabby male and a dilute torbie female. The term "dilute" refers to the soft peach, gray, and cream coloring, rather than the more typical orange, black, and white. "Torbie" refers to a combination of tortoiseshell and tabby markings. Unlike most tortoiseshells, however, this little one was very timid and shy. We named them Riley and Katy.

Riley was bold, inquisitive, talkative, and demanding. He was also affectionate and had a loud purr. He loved to eat and could smell food while in a deep sleep on the other side of the house. He would taste anything twice, but would not, under any circumstances, share the food bowl with his sister. He also wanted very badly to go outside.

Katy, besides being shy, was hesitant, quiet, and demure. She had a gentle purr. She was playful and loved to chase and climb. She was a very sound sleeper.

Once they were allowed outside they lost no time in exploring, hunting, and chasing one another. They also liked to climb trees. The only problem with this was that they often climbed too high to get down. They also managed to get onto the roof of the house. We could hear them yowling pathetically from either place. I would do the tree rescues while Dan would get them off the roof. Eventually we left them stuck until they finally figured their own way down.

Riley seemed to like Rascal more than his own sister, and Rascal barely tolerated Riley, but for the most part Riley and Katy left Rascal alone. This was good because eventually his remission was over and the last thing he felt like doing was putting up with a couple of enthusiastic young cats.

CROSSING THE RAINBOW BRIDGE

Watching a beloved pet slowly die is a difficult thing. If they are comfortable and not suffering, then a natural death is in order. If they are suffering or having other difficulties, then the question of whether and when to put them down is a looming one and not easy to answer.

Rascal had beaten his cancer for several years, for which we were exceedingly grateful. Eventually, we knew something was wrong once again, because he began to slow down and lose weight. A visit to his vet confirmed that his remission was over. Unfortunately the cancer cells that survived the first rounds of chemicals were now resistant to them and would be more difficult to beat. Since we had initiated treatment with the best course of action, the next step would have been to proceed with the second best treatment. If he survived that, then the third.

We had agreed to surgery and chemotherapy the first time because we were not ready to let him go. Thankfully we had the funds. He was in remission when we moved to the homestead, and both Dan and I hoped he had beaten the cancer for good. Having it recur was a disappointment, but not entirely unexpected. Because of that we were somewhat emotionally prepared, knowing that the chances of extending Rascal's life much longer were getting slimmer. We had a decision to make and had to weigh all the factors.

Some folks think that life at any cost for a pet is worth it. The tendency is to think of cost in terms of money, but there are other costs as well, emotional costs, for example. *I* might feel better about extending my animal's life, but for a cat who lived to hunt and chase, the loss of freedom is a huge price to pay for my emotional happiness. Somehow that felt selfish.

For Rascal, one of the considerations was the numerous lengthy trips to the vet which would be required for the next round of treatment. If Rascal hated anything, it was riding in the car. We were getting to where the medications were too toxic to be administered at home; the vet would have to do this. In addition, Rascal would need frequent blood tests to monitor his liver and kidneys. Could we honestly subject him to all this in hopes of a few more months of life? In the end we decided that we couldn't. We would let him live as long as he was comfortable and could do the things he loved to do.

For the next couple of months Rascal had good days and bad ones. On the good days he would go outside and make his rounds. He knew where the mice and chipmunks hung out, so he could usually be found there, keenly but patiently watching for movement. He knew that the comfiest place to take a nap was in the hay. He remained faithful to accompany me when I went out to the garden, and he would keep me company while I did various gardening chores. His annual check-up was coming up in August and on his good days I hoped he could make it for that.

On bad days he didn't want to go out, didn't want to eat, didn't want to be bothered. On days like that we'd debate whether or not to take him to see his vet, Dr. Cal. We knew that in all likelihood, if we took him, he probably wouldn't come home again. It was hard to be brave enough to face that. Then in a day or two he'd rally, although we wondered how he kept going. He was all skin and bones and very fussy about food. In fact, he would drive me nuts with his "What else have you got?" look each time I set down another dish containing a different tidbit. As long as his good days outnumbered his bad, we'd hang in there with him. I hoped that when his time came he would fall asleep peacefully and never wake up. If he did need help crossing "The Rainbow Bridge,"* we just hoped we'd know when he was ready.

One morning Rascal woke me up around 6:30 and asked for breakfast. He'd been very hungry lately, but his favorite dish seemed to be raw chipmunk, which of course I didn't have; he caught that himself. That morning he ate a big helping of chopped raw chicken breast, and then accompanied me outside while I worked in the garden. He made his usual rounds, and when I went in for breakfast, he came in too and took a long nap. After lunch I came in from checking on the chickens and goats, and found him lying on the kitchen floor, breathing with great difficulty. I called Dan to let him know what was up, and then made an appointment with the vet.

We had several hours before his appointment. I spent the time with him, talking to him, using all the words he knew, and telling him how much we loved him. We've tried to keep an emotional distance from our other critters but not our cats. Our cats are contributors to our lifestyle, but they are also pets. We've tried to learn not to get too attached to the goats and chickens, but the cats, and especially Rascal, find places in our hearts from day one.

Dr. Cal said his lungs were filling with fluid and that I had done the right thing to bring him in. He would have suffered terribly as his breathing got worse, eventually drowning in his own fluids. We were able to ease that suffering by helping him to the place of no suffering.

Cats have come and gone on the homestead, but Rascal remains the standard of cats. Perhaps that's because we had him for ten years, longer than we've had any other. Or perhaps it was because he was the most unique cat we've ever known.

---

*"Crossing the Rainbow Bridge" is a phrase sometimes used by pet owners to signify a pet's death.

ODE TO RASCAL

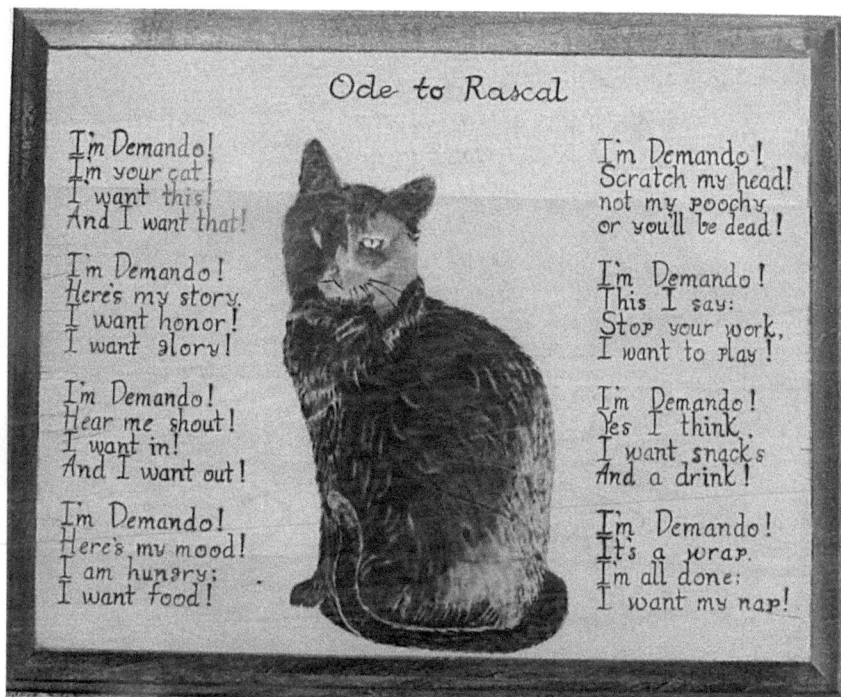

Ode to Rascal

I'm Demando!
I'm your cat!
I want this!
And I want that!

I'm Demando!
Here's my story.
I want honor!
I want glory!

I'm Demando!
Hear me shout!
I want in!
And I want out!

I'm Demando!
Here's my mood!
I am hungry;
I want food!

I'm Demando!
Scratch my head!
not my poochy*
or you'll be dead!

I'm Demando!
This I say:
Stop your work,
I want to play!

I'm Demando!
Yes I think
I want snacks
And a drink!

I'm Demando!
It's a wrap.
I'm all done;
I want my nap!

By   Daniel Tate
Leigh Tate
Heather East
Nolan East

---

*"Poochy" was what the family called Rascal's belly. He considered it off limits, and the combination of the word plus a hand approaching his belly was an occasion for teeth and claws.

POISONED

Riley and Katy continued to grow and began to develop very different personalities. Riley was the bolder of the two, while Katy remained a scaredy-cat. She was also a lap cat which was a treat for me, because Rascal had rarely consented to lap time. Even though they were litter mates, Riley really didn't seem to have much time for Katy, unless it was to take over her food bowl or her nap spot. Her favorite nap spot was a basket that I had filled with old towels and topped with a hand-crocheted, hand-felted kitty bed. Riley's favorite trick seemed to be to roust her out of it.

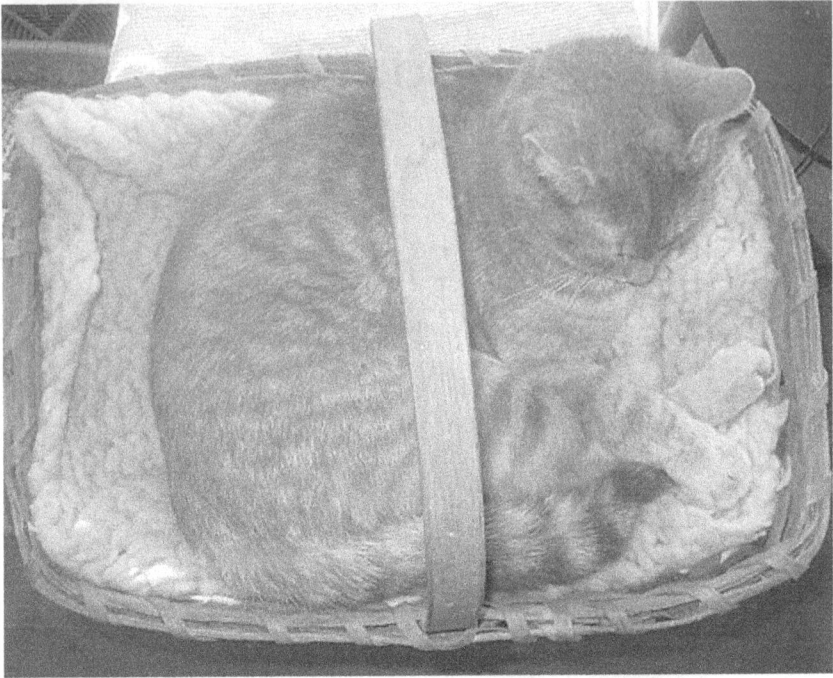

Setting: Katy has the basket.

Problem: Riley wants the basket.

Solution: Rouse her under the guise of giving her a bath.

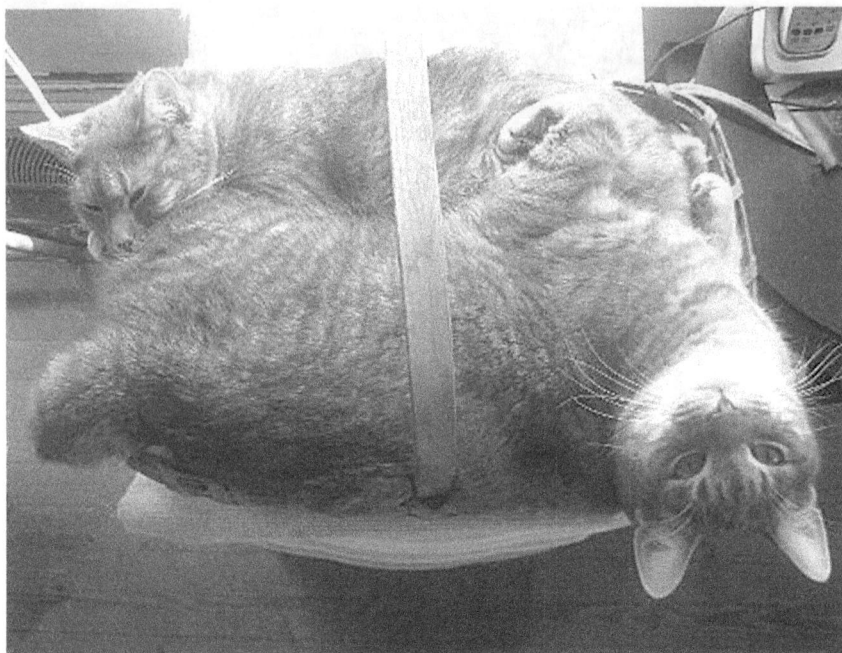

Moderately Acceptable Outcome: Kitties share the basket.

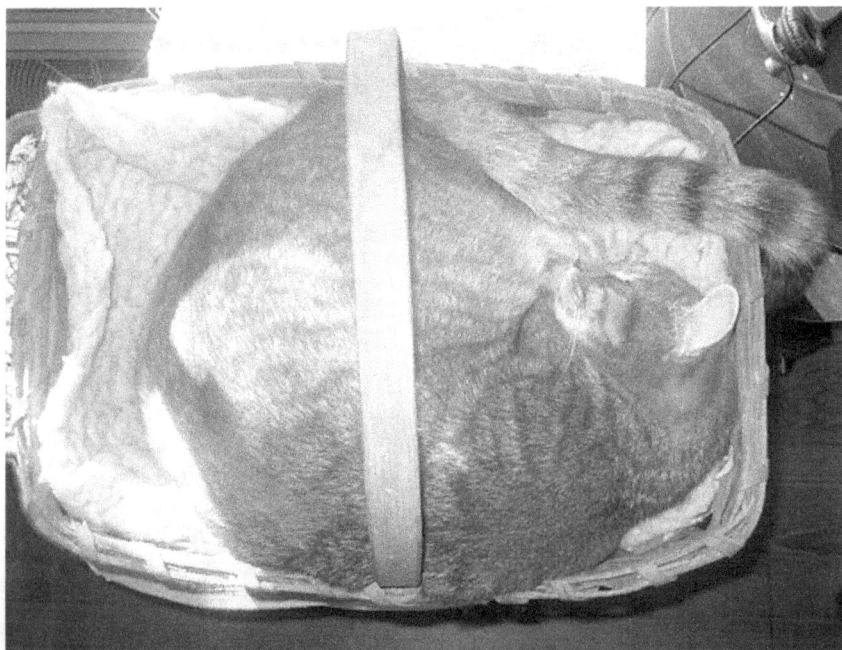

Preferred Outcome: Riley gets the basket all to himself (and he usually did).

One morning Katy laid tightly curled up in her kitty basket and did not get up. I discovered she was burning up with fever, which alarmed me. I debated what to do. The first thing that came to mind was taking her to the vet. But I also treat some problems myself, having had good results with garlic, even though garlic is trickier to give to cats than to goats, because there is no form I've found yet by which they will take it willingly. With Rascal, I finally learned to cut a small sliver and dip it into strained meat baby food. I still had to cram it down his throat, but he only tasted the meat and would often beg for more. It got to where anytime he smelled garlic he would come beg for it, because he associated it with the tasty meat.

As an antibiotic and antiviral, garlic works very well and is safe for human consumption, but I have since learned that it can cause hemolytic anemia in some animals, particularly dogs, cats, and horses. The disulfides in the garlic will rupture the oxygen-carrying red blood cells of these animals. The result is a condition called Heinz body hemolytic anemia; the Heinz bodies being the fragments of the damaged red blood cells.[1] Consequently, I am not recommending its use as a treatment for any critter, I am simply sharing my thought process as I tried to analyze the problem.

Even so, I chose not to use a home treatment on Katy, but to take her to the vet. He diagnosed enteritis and gave her two injections of antibiotics. He told me she'd be back to normal within a day or so. I took her home.

She was not better by the next day, nor the next. She rallied a little so I took her outside for fresh air and sunshine. She was disoriented and did not seem to know where she was or how to get back to the house. I had to fetch her and bring her back inside. She became weaker, neither wanting to eat nor get out of her basket. There was a personality change as well; she wasn't my Katy. Something had changed. If I had been worried before, I was really worried now. I took her back to the vet.

This time Dr. Mac did blood work. It indicated that her creatinine level was off the chart. Creatinine blood level is an indicator of kidney function, and her levels indicated renal failure. For that, there is no cure. Considering her young age, he thought poisoning was the most likely cause, but this made no sense. We keep no poisons in our home or on our property. Things like antifreeze for the vehicles are kept in their original containers in our vehicles. I could not see any way Katy could have ingested poison. Even if she could have gotten into something at a neighbor's she was too weak to wander that far on the day she was allowed outside.

The kindest thing would have been to have her put to sleep right there at that second vet visit, but I wasn't ready to let her go. I brought her home and tried to care for her there, while I researched this mysterious turn of events. In researching the antibiotics given I discovered that a side effect of one of them is kidney failure. As I write this years later, I cannot remember which drug it was, but it is commonly given for enteritis in both pets and humans. The risk of this side effect is equal for both, and my Katy was unfortunate enough to be one where the risk outweighs the benefit.

Her weakness and disorientation were the results of being poisoned by her own natural metabolic waste. As desperately as I wanted to save this little cat, her personality was gone before her body was. It was heartbreaking to see her wasting away, so Dan took her back to the vet's office to be put to sleep.

Is it possible that the enteritis had been caused by poisoning so that her subsequent problems and demise were from that? That scenario is certainly possible, although I tend to disbelieve it. Why? Because metabolic conditions usually cause body temperature to decrease, while infections increase it. Katy's high fever is my biggest clue to what I think happened.

We buried her next to Rascal. She was the third cat we'd lost in our first year-and-a-half on the homestead, and the numbers were devastating. At least they were to Dan and me. Riley seemed to take no notice. For him it was cat business as usual, and he seemed to like not having competition for our attention. Riley liked being an only cat.

Expanding the Rodent Control Department

Riley was a pretty good mouser, but after we lost all the baby chicks to the rat, we decided it was time to expand the rodent control department. As much as Riley liked being an only cat, there were more important things at stake than his personal preferences. There were certainly enough rats, mice, shrews, moles, rabbits, chipmunks, snakes, and squirrels to keep more than one cat busy. I looked at free kittens on Craigslist, but never reached a decision. When August finally rolled around, it occurred to me to go back to our area animal shelter. August was the month I had adopted Catzee, the month they discounted adoption fees for cats. I headed on over.

The shelter was busy that day and there were three crowded rooms of cats to see. We figured Riley would be less threatened by a young female than another male, one old enough to hold her own. That seemed the best option for a peaceful transition. I spent time playing with and looking over each potential candidate. A few I disqualified for various reasons, but of the remaining choices, none seemed quite right.

All the kittens were active and I couldn't resist watching them. One pair was particularly cute. It consisted of a calico-tabby female and a white and ginger tabby male. The little girl cat was quite a lover, begging with her loud, affectionate purr to be petted and picked up. The little boy cat, on the other hand, ducked and tried to hide in the back corner of the cage. He hissed at anyone who tried to touch him. Not being one to take such things personally, I picked him up anyway. He didn't growl, bite, or try to scratch, he was just scared. Once in my lap, he settled down and tried to scrunch himself up into the smallest size possible.

"I'll take them both," I told the volunteer. "Great!" she replied with a huge smile. As we filled out the paperwork and I paid the fees, she told me she was so relieved that I was taking them both. They feared the little female would be adopted without her brother, and that he would end up being classified as unadoptable because he was too frightened to be friendly.

I brought them home and set up a temporary "new kitty introduction room" in the pantry. I put a baby gate in the pantry

Sam and Katy. Katy was bolder, but Sam was more energetic.

doorway to keep them in. This would give them some territory to get used to, plus, I hoped, let Riley get used to the idea of his new housemates before being overrun with kittens.

Little Girl Cat was happy, friendly, and constantly purring; instant cuddle cat, just add lap. She immediately made herself at home. Little Boy Cat was timid and shy. He was slow to coax out of the kitty carrier. If I tried to pick him up he gave me a baby hiss and tried to run away.

As I predicted, Riley was not pleased. He quickly hopped over the gate to investigate. He didn't stay long because both kittens were fascinated with him, apparently immune to his hissing. Introductions concluded with both kittens sitting on their side of the baby gate watching Riley, while he lay on the kitchen floor tossing hisses at them over his shoulder.

We named the little male Sammy. I noticed that if I started to pet him from the back he was okay, but if he saw my hand coming he would flinch and try to run away. It took a long while before he would purr for me. To this day he does not like hands and prefers not to be petted or picked up. He has turned out to be a good hunter—catching chipmunks, baby rabbits, squirrels, and rocks. For some reason he likes to play with rocks.

Little Girl Cat took longer to name because nothing seemed quite right. I found myself calling her Katy, but we'd already had a Katy. Besides, this kitten was nothing like our first Katy. Our first Katy was the shy one, but this little cat was happy, friendly, and constantly purring. I tried to think of another name, but would refer to her as Katy if I didn't catch myself. She is Katy to this day.

Katy, for all her bravado, is more timid than the boys when she is outdoors. She limits her territory to the garden, goat shed, and blueberry bush. She'd rather hang out in the house. That's okay, really, because there continues to be mouse activity in the attic, despite our best efforts. There are rodents enough to go around.

Bugs, mice, shrews, feathers, carpenter bees, and butterflies are the most common victims of Katy's hunting prowess. Unfortunately, she also caught birds. This much-criticized habit amongst cats was particularly bad in spring, when nestlings abound. We indulged her tendency to want to stay indoors during this time of year, to curb casualties. She was content to chatter at birds through the windows. Fortunately, she outgrew this.

For others who want to deter cats which particularly like to catch birds, I have a few observations to share. My cats will only jump our welded wire fences where there are wooden brace posts to climb. They don't climb any of the metal uprights that we use in our fences. Fences secured with t-posts or rebar work well for protecting areas where birds nest or feed. While some bird lovers would say this is still not acceptable, for us it is, because we need the cats to control rodents. The damage rats do to our homestead is not acceptable. Nor is an increased snake population from not controlling the mice and rats. Those same snakes will eat our chickens' eggs and baby chicks in addition to small rodents. Snakes also raid song bird nests to hunt for eggs, baby birds, or mama birds setting on a nest. Some folks may prefer snakes to cats, but I can't imagine a snake doing double duty as a lap warmer.

In nature we hope for balance. Sometimes the pendulum swings one way, favoring the predators, sometimes the other, favoring the prey. Because many of our domesticated animals are prey, our predatory cats are a necessary part of a complex equation.

The Kitty Door

Having to play doorman to a cat can be frustrating, especially considering how fickle cats can be. Rascal used to want us to open the door just so he could look out. In fact, he did not like closed doors and asked that every door be kept open. Others of our cats have been more typical, requesting to go out but deciding to take a bath or a lick before proceeding. Or more annoyingly, deciding not to go out once we've opened the door for them. To minimize our roles as doormen, we decided to add two kitty doors when we did our kitchen and back porch remodel: one from the kitchen to the back porch, and one from the back porch to outside.

The back-porch-to-outside kitty door went in first because we worked on the back porch first, planning to use it for temporary cooking and clean-up while we gutted and rebuilt the kitchen. Our plans for the second little door were foiled by the construction of our 90-something-year-old house; a 4x4 post was placed right where the kitty door should go. Never mind, we could let them out the kitchen door and they could let themselves outside.

I hung bells on the kitchen doorknob as a kitty door bell. They quickly learned to bat the bells anytime they wanted to come in the house. This scheme worked very well most of the year, but we began to have problems when the weather was mild and I left the kitchen door open.

One beautiful October day, when Riley was still an only cat, I was in the pantry filling the water pitcher from our Berkey filter. The kitchen door was open to the back porch because the weather was warm and pleasant. As I stood there waiting for the pitcher to fill, I heard the kitty door.

"Riley, is that you?" I asked, not expecting it to be anyone else.

Next, I heard loud crashing

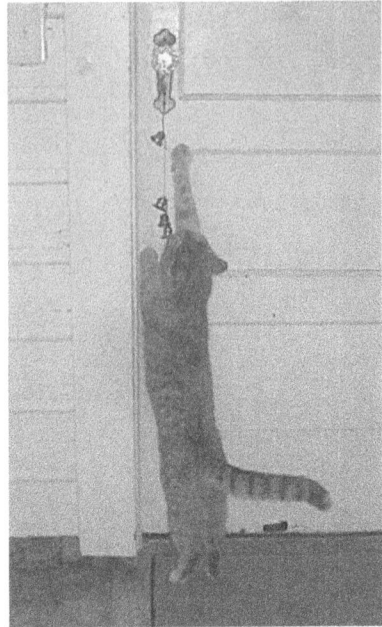

Our first Katy demonstrates how to use the kitty door bell.

and banging. That alarmed me, but the pitcher wasn't full yet so I couldn't go see what it was. Suddenly a chipmunk came hightailing it around the wood cookstove toward the pantry, and me. Riley was right on its tail.

I screamed, the chipmunk panicked, but Riley never missed a beat. After a flurried turn or two around the kitchen, the chipmunk made a dive for the bathroom with Riley right behind. I quickly closed the door, relieved that I wouldn't have to chase a chipmunk around the house and try to catch it.

I had to leave the kitchen after that, because I couldn't listen to the life and death game being played in the bathroom. Dan and I have a policy of non-interference when it comes to cats and their business, and this was Riley's business. I just wished he'd do his job outdoors.

Now, I'm not one to assign human thinking to animals. Animals can reason in their own fashion, and some can be pretty smart about it. I think Riley's that way; I think Riley is a smart cat. Just a day or two prior, I saw him in the driveway with a chipmunk he'd caught. He was playing "cat and mouse," where he pretends his attention is elsewhere, until his "mouse" makes a run for it. (It seems a cruel game, I know, but that's what cats do.) That particular 'munk made a run for it and ran right up a tree. Call me crazy, but I think Riley brought this one into the house so it couldn't get away!

This tale does have a happy ending. Once Riley finally asked to come out of the bathroom, I assumed the chipmunk was dead. It certainly looked dead. When Dan went to pick it up, however, it was very much alive—a bit stunned, but alive. He set it by one of our numerous chipmunk holes outside and it disappeared down the hole in a wink.

After that I kept the kitchen door closed, and we'd peek out the kitchen door window when Riley rang the kitty doorbell to come into the house. We told him he was only allowed to come in if he was alone. If he had something in his mouth, he was told to take it back outside. This, he would not do, and we had to catch chipmunks and mice on the back porch ourselves. What a nuisance!

This trick wasn't particular to Riley. When Sam and Katy became old enough to hunt, they would bring catches onto the back porch as well, or worse, into the house if the kitchen door was open: chipmunks, mice, birds, lizards, cicadas, even a baby bunny. At one point, a "catch" hid in the house so well that none of us could find it, until the telltale odor of expiration made me ransack the place to remove it. On the bright side, this episode did cause me to completely clean and reorganize our storage room.

In the end, we saw only one solution for this ongoing situation. Keeping the kitchen door closed didn't prevent them from bringing their prizes onto the back porch, plus, I didn't like closing up the kitchen on beautiful breezy days. The only other thing to do was to set the kitty door to one way—out only. That's what we did. I moved the kitty doorbells to the screen door on the back porch, but for whatever reason, the cats now refuse to use them. As her way of letting us know she wants in, Katy will catch the screen door with her claws, causing it to bang. Sam will just sit there and wait until we notice. Riley now prefers the front door, unless someone happens to be handy in the back when he wants in.

So, the kitty door was a good idea and remains somewhat convenient, but we still play kitty doorman. Such is life with cats.

To Mom, a Message from Your Cat

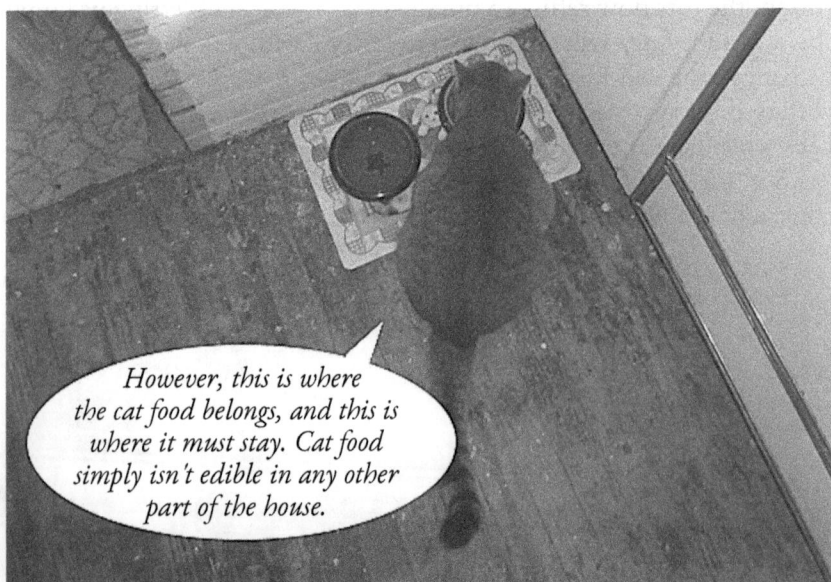

YOU'RE GOING TO BE MY FRIEND, LIKE IT OR NOT

Sammy was infatuated with Riley from the first day he saw him. He followed him around, went where he went, did what he did. When Riley went to eat, so did Sam. When Riley took a nap, so did Sam. When Riley went out, so did Sam. Wherever Riley was, Sam wanted to be also. Riley, however, made it very clear he didn't want Sammy following him around. Sam would make such a pest of himself that we'd sometimes keep him inside when Riley went out, just to give Riley a break. We told Riley he ought to be flattered to be admired so much, but Riley was not impressed. He would hiss, growl, and swat at Sam to deter him. Sam was rarely deterred.

One day Riley was asleep on the bed when Sammy jumped up. Sam lay a respectful distance from Riley and appeared to take a cat nap. By the time Riley was sound asleep, Sammy had inched his way over so that he could sleep snuggled up next to Riley. That became Sam's naptime routine. It worked pretty well until Riley woke up. Then it was hisses and growls all over again.

Eventually it got to where Riley might stir briefly with Sam at his side, but would ignore him. If he was particularly comfortable when Sam arrived, he pretended not to notice. He began to let Sammy give him a bath, and Sam seemed to become his permanent sleeping companion.

Encouraged by his progress, Sam next attempted to engage Riley in play. Riley, being older and more dignified, was not willing to participate. Riley would play with Dan or me, if one of us got out a string or the laser light. But if Sam or Katy showed up and wanted to join in, Riley would immediately quit playing and walk away.

Sammy's pouncing and ambushing seemed to irk Riley more than the sleeping situation. A couple of times he really let Sammy have it, giving him a good trouncing for attempting to play. That was the first time I'd ever seen a cat get his feelings hurt. Sam would come in the house and mope for days whenever Riley rejected him like that. But he didn't give up.

It took several months, but eventually I saw Riley and Sam racing and chasing each other. It finally got to where Riley would initiate the fun by ambushing Sam, and then the chase was on. Sam still tends to push his limits, however, and on occasion there will be a tail-whipping, caterwauling standoff. At those times I figure Riley has had enough.

In the end Sam's persistence paid off, although that doesn't mean his feelings are 100% reciprocated. Riley will, at times, become annoyed and hiss him off. Or sometimes Riley just gives him the slip to have some personal time for himself. Sam will still try to hunt him down, but he's finally learned to back off when warned to do so.

Where is Katy in all this? She has never shared Sam's enthusiasm for Riley, and Riley's toleration of Sam does not extend to her. She keeps her distance. She can be a pest in her own right, but she always goes for Sam and never Riley. She often initiates the play, which is fine, until Sam gets too wound up. Once he gets going he enters the "Sammy Zone" and doesn't let up. This invariably concludes in hissing, growling, swatting, and tearing around the house. More than once I've had to rescue Katy from Sam's wild abandon, always pointing out that she usually started it.

It's interesting to see how they've worked out their own social order, sometimes with respect for one another, sometimes not. Although they don't work as a team, they do a good job of keeping the homestead rodent population under control. That's what they're here for, and that's all we ask. Throw in some feline affection and lap-warming during winter, and that's our life with cats.

OF CATS AND HOMESTEAD SUSTAINABILITY

In each section of my tales I've talked about how we are working toward a sustainable relationship with a particular critter: the steps we've taken, the challenges we face, and the potential for success. With the cats (and the dogs when we had them) there has been a lot of compromise, with commercially-produced pet kibble still on the menu. While I seek to avoid potential genetically modified ingredients (soy and corn), I still succumb to the convenience (and the expense) of store-bought cat food.

In researching natural diets for cats and dogs, I discovered two camps, the Raw Meaty Bone proponents and the BARF (Biologically Appropriate Raw Food) advocates. I also discovered there is quite a bit of competition—and argument—between these groups. Rather than try to decide for myself which one is "best," it makes more sense for me to understand the basis of both and apply what I can to our situation. Understanding the nutritional benefits of these diets is helpful, while arguments about what constitutes "natural," evolutionary rationales, and whether or not dogs and cats are omnivorous versus carnivorous, are not. I am much more concerned with being able to feed all our animals from what we can raise on the homestead or forage locally, than I am with developing an adherence to any particular theory of diet.

What progress have we made in feeding the cats from the homestead? They can, and do, catch quite a bit to eat on their own. I do not leave cat food outdoors for our cats; they must come into the kitchen if they want it. I do not feed canned cat food, only dry. During summer, Riley and Sam rarely come into the house. If they do and run right to the food bowl, we have a pretty good idea that hunting was unsuccessful.

Other things I feed are raw eggs (beaten) and fresh milk. We've learned that cats, dogs, chickens, and pigs enable us to have zero waste when we butcher animals for meat. They will eat what we will not, including the raw bones. In addition to the organ meats, the cats like bone broth and most soups and stews. Dogs too, love leftovers. When I defrost a roast I drain off any blood and give it to Riley. He loves it, although the other two just sniff at it, which points to a well-known problem when it comes to feeding cats—they are picky. I will actually offer them anything they are interested in. Some they'll sample, some they'll actually eat, some they'll walk away from.

If we ever reach a time when there is no more commercial dry cat food, I'm sure there will be much complaining and protesting. But even a cat will eventually succumb to reality. If it comes to that they'll have to either eat whatever leftovers or scraps I can offer, or go catch their own, which is why they're here in the first place.

Pig Tales

Considering Pigs

Getting pigs had been on our annual list of homestead goals for several years. Every January 1st, there it was on the list, "get pigs." When our annual evaluation rolled around in December, it was always one thing that did not get checked off the list; rather it was transferred to the goals for the upcoming year.

With our primary goal of food self-sufficiency, pigs seemed a logical choice for us. We'd read Laura Ingalls Wilder's *Little House in the Big Woods* enough times to know that they could supply us with quite a bit to eat, both meat and lard. What meat eater doesn't love sausage, bacon, and ham? What pastry lover doesn't adore flaky, melt-in-your-mouth pie crusts? In addition, pigs could fit quite nicely into the sustainability of our homestead. They will eat food scraps that other animals won't, add manure, of course, and will work for their food in ways other animals can't because of their natural tendency to root. I'd read somewhere that they will eat blackberry and morning glory roots (two big problem weeds for us), and I figured that since the

Japanese have recipes for kudzu roots, then they must be edible too. I hoped pigs would agree. Doesn't that sound like the perfect homestead animal?

We'd need to keep a breeding pair for food sustainability, but considering our lack of experience with pigs, I also figured it would be best to start out with a "practice" pig or two. We could raise them in order to learn all about pigs and then use the meat.

I set about trying to find my practice pigs by browsing Craigslist from time to time. This never lasted long because I was always dismayed at the price. Perhaps I was looking at the wrong time of year, but at the time it seemed the "going rate" for mixed breed weaner pigs ran into the hundreds of dollars each. The cost of buying, fattening, slaughtering, and butchering one of these pigs didn't seem like an economical way to have pork in the freezer.

I had heard horror stories about aggressive pigs, so I was wary of certain breeds. I really wanted a heritage breed, and initially considered Red Wattle Hogs. According to the The Livestock Conservancy they are mild natured, adaptable, excellent foragers, show a good growth rate, and are a good choice for the novice pig farmer. The sows are good mothers and produce large litters of ten to fifteen piglets. They also grow to somewhere around 600 to 800 pounds, or larger.[1] That's pretty big and I wasn't sure how well we could feed a couple of pigs that size with our small acreage. Still, I started looking. Like Kinder goats, however, there were none to be had anywhere near me nor in neighboring states.

When I mentioned my pig hunting on my blog I started to get suggestions from readers about American Guinea Hogs. This is a true American heritage breed, originally raised as a lard pig right here in the southeastern United States. Like the Red Wattle, the breed is listed as "threatened" by The Livestock Conservancy. They sounded like the ideal homestead pig: small (adult weights reaching 150 to 250 pounds), docile, friendly, and excellent foragers. They earned the name "yard pigs" because at one time they were kept in the farmyard to catch and eat snakes and mice. They are described as "hardy and efficient, gaining well on the roughest of forage and producing the hams, bacon, and lard essential for subsistence farming."[2] The American Guinea Hog was exactly what I was looking for.

We wanted to make sure we could care for them properly, so the next step was to research their basic needs for food, fencing, and shelter. I found a lot of good information at the American Guinea Hog Association website.[3] I learned that our existing welded wire fence would be suitable, that a small portable shelter would be fine, and that I would need to provide plenty of clean water, a mud wallow, and shade in hot weather. I also learned that these pigs are a grazing breed which means they do very well with pasture grasses, forage from wooded areas, and occasional garden and table scraps.

The feeding part was especially good news, because sources for fattening hogs for meat indicated they would go through hog feed by the barrelful. This had worried me. With our self-sufficiency goal we needed to be able to raise everything our pigs would need and not rely heavily on store-bought feed. Standard (commercial) breed pigs seemed to require lots of feed in the weight gaining stage. The American Guinea Hog, however, did not have this requirement. In fact, according to the Quartz Ridge Ranch website, care must be taken not to overfeed this breed. As a lard breed they can easily get too fat.[4]

That information pretty much sealed my decision to go with this breed of pig. I'd had a bountiful crop of goat kids that year and sold quite a few. With that money I could buy my piglets. I began to keep an eye on Craigslist and before long, I saw an ad for five-week-old American Guinea Hogs. I'd hoped to buy a breeding pair, but all of this breeder's piglets were related so I'd have to find the second pig elsewhere. No matter, we were on our way to becoming the proud possessors of our first pig.

WALDO

We brought him home in a pet carrier, the first of our two American Guinea Hogs. We let him out near the goat shelter in our front pasture, where he immediately attracted the attention of our bucks, chickens, and cats. None of the other critters knew what to make of him. He came from a farm where the pigs mingled with Nigerian Dwarf goats, so he was at home with my billy boys. They needed a bit more time to get used to him, though. We named our new little pig Waldo.

Waldo spent his first day looking for someone or something familiar. He must have run the perimeter of the fence at least a dozen times. Realizing he could easily slip under the gates, we quickly blocked them with logs and rocks.

After the first day he settled down to his piggy business. He was a very busy little fellow, always on the move, always eating, always grunting with an occasional squeal and rare bark. He was shy but would come sniff my hand and receive a scratch on the head. He quickly learned to come running when he saw me with his dish of table scraps and whey in the morning. He adored the whey (still does) and slurped it right down. He would burrow in the straw at night between

the hay feeder and the goats' sleeping pallets. We tried to make him a mud hole but he seemed to prefer his water dish for cooling off. He made himself right at home.

One day I heard the goat kids hollering. Initially I didn't think much of it because kids will holler about everything: they can't find their siblings, they can't see their mom, they can see their mom but she's on the other side of the pasture (so rather than going to her they holler for her to come to them), or simply because somebody else starts hollering. Zoey's Li'l Red was getting pretty worked up, however, so I thought I'd better go check. As I approached the gate, all nine goats ran past. Something must be going on. As I got closer, I saw Waldo! How did he get in with the does and kids?

Waldo came running up to greet me, apparently very pleased with himself for finding his way into the back pasture. A quick walk of the fence and a gate check revealed that somebody (or some pig) had pushed his way through the log barricade and wiggled under the gate. Pigs are notorious escape artists and obviously our gating situation wasn't pig proof. Actually it wasn't chicken or baby goat proof, either. All will squeeze their way under gates at any chance they can get.

No pig wrangling took place, but after Waldo made his way back, I added a few more rocks to the under-gate log barricade. I could see why he liked the doe pasture. It had quite a bit of clover and pigs love clover. No harm was done, however, other than startling the girls and their kids, who all made a great show of being alarmed. I figured they'd eventually get used to seeing him.

Pigs are companionable animals and not loners, so it is better to have at least two. I've been told that for those being raised for meat, having competition for the food makes them gain weight more quickly. Waldo was for breeding, however, so speedy weight gain was not a priority. However, one plus zero does not equal more pigs! He needed a sweetheart and my search for one was soon rewarded. I quickly put a deposit on a little gilt (female pig that hasn't given birth or "farrowed"). She would be ready for pick up in a few more weeks.

POLLY

We brought home Waldo's bride-to-be in the same pet carrier we used to transport Waldo. She was a two-month-old registered American Guinea Hog, shy and a bit skittish, with long eyelashes and a long snout. I named her Polly.

She spent her first day running the perimeter of the fence as Waldo had done, looking for a way back to familiar things. I made sure the gates were well blocked this time, but that didn't stop her from trying to get through them. By now the bucks were used to pigs and didn't think much of her. To introduce her to Waldo, I put her on the other side of the fence, to let them smell and see one another first.

What did Waldo think? Meh. They gave one another an occasional sniff through the fence but each was more interested in hunting for acorns. Even after I opened the gate to let them in together they didn't pay much attention to one another. Eventually that changed, of course, and soon Polly began to follow Waldo around.

So now we had our foundation breeding stock, destined, through their progeny, to provide us with bacon, sausage, pork, lard, and more pigs. Our homestead was beginning to take shape.

"They Won't Tear Up Your Yard"

"And another thing, they won't tear up your yard."

When the seller mentioned this I almost changed my mind. Here we were in her pig pasture, surrounded by more than a dozen five-week-old American Guinea Hog piglets, their mothers, several dozen chickens, and four or five curious Nigerian Dwarf goats. There was no sign of rooting anywhere. Rooting was one of the reasons we wanted pigs! However, I had already handed her the $150 and they were in the process of catching my newly procured baby pig. It seemed a bit too late to change my mind.

We have five paddocks, ranging in size from a quarter-acre to an acre. The plan has been to work on pasture improvement and maintenance, one paddock each year. The pigs were to play an important part in that plan because of their natural tendency to root through the soil with their snouts. They are natural tillers of the soil this way, and they will eat things we'd like to get rid of, things like blackberry, morning glory, squirrel-buried acorns, and kudzu roots. In my work-smarter-not-harder scheme the pigs and chickens would be given the field needing the most improvement. They would till, clear, and fertilize the soil by their natural behaviors. Then I would test the soil to remineralize as needed, and we'd plant whatever crops we planned to grow for the year. After harvest the goats would be let in first to browse on the harvest remnants, followed by the pigs and chickens once again. Lastly we'd reseed for pasture forage for the next several years. If my American Guinea Hogs didn't root, however, this plan was pretty much crippled.

Imagine my delight when one day I went to check on the pigs and discovered a number of pig-created furrows in the ground. They indeed would "tear up my yard!" As Dan and I stood at the fence watching them, we pondered why Waldo's breeder's experience was different than ours. We reckoned the reason was because her soil was so compacted and trampled from the number of animals kept in the paddock. It was hard and there wasn't much growing. Our soil, on the other hand, was better aerated and full of a variety of plants.

We had a dry spell that summer. The does had finally gotten used to the pigs, and I let the pigs graze that pasture because they loved the clover and ate the grasses as well. With no rain, however, the clover and grasses began to dry up. I was dismayed when ground ivy took

We have learned that our pigs are less likely to root when the ground is dry, or if there is plenty of pasture for grazing and acorns to hunt and eat.

advantage of the condition and overran my lovely pasture. Goats won't eat ground ivy, which smothers out other plants with its dense root system. The pasture we had worked so hard to establish was in danger of being lost.

The pigs did not eat the ground ivy either, but they did go after whatever was under it. They'd get their snouts under its shallow roots and roll it up so that it looked like rows of bunting to hang on a balcony as a parade decoration. I discovered that while ground ivy is difficult to pull out by hand, once the pigs uprooted it, it was easy to remove. It wasn't long before they uncovered a whole lot of soil.

They did such a good job of tilling the soil that it was not going to require plowing or tilling on our part. Dan used the rake that came with his walk-behind tractor to push the dead ground ivy into great piles, which I loaded into the wheelbarrow and dumped in the chicken yard. The chickens were more than happy to scratch through it in search of anything they might fancy. All that was left to do was to broadcast a mix of pasture and forage seed for a fall and winter planting. I did this by hand and we moved the pigs into the adjacent field with access to the woods.

Some farmers don't want their pigs to root. Those raising pastured pork, for example, want their stock to eat the pasture, not root it up. The recommended solution to this problem is to feed them more. That might make sense for their situation and goals. I'm guessing Waldo's breeder kept her pigs well fed so that they had no incentive to root. For us, it makes more sense to let them forage for their own food by rooting. We find that the pigs are very happy to be doing this and by no means look undernourished. By letting them root in areas that need it, we are partnering with them for the sake of the homestead. This is working smarter, not harder, at its finest.

MYSTERY OF THE TIPPING BUCKET

I have two kinds of water buckets: big ones for the tall goats, and smaller ones for the shorter goats. I check them all several times a day, cleaning out and filling as needed. I became puzzled when I kept finding Pygmy buck Gruffy's small bucket knocked over and all the water spilled out. If he'd been with the other bucks that would explain it; they often knock things over in their sparring over which is the most macho he-goat. But it was breeding season and the bucks were in rut. Gruffy was now by himself in the buck pasture precisely because of all the sparring. As the only Pygmy goat and shortest buck, I felt he got picked on too much and I wanted to give him a break. He could hang out with the other bucks through the fence, but the only other critters with him in the buck pasture were the chickens and the pigs.

Several times I refilled that bucket and each time I went back to check I found it knocked over again. It wasn't until I saw Waldo bumping it that I realized he was the one knocking it over. Was he having trouble getting his own drink? I kept a pan of water nearby for the pigs so I didn't think he could be thirsty. Maybe he was just clumsy? I decided I'd better secure Gruffy's bucket so that it couldn't

I hoped that securing the handle would keep the bucket from being tipped over.

be tipped over. I found a hook and installed it in the buck barn at Pygmy goat height. Hopefully that would solve the problem.

I brought a clean bucket of water and Waldo came running over. He immediately started pushing on it. Suddenly, I realized what he was wanting. I hadn't refilled his piggy pool. I had cleaned it out during an autumn-like cool spell earlier in the month and forgotten to refill it when the weather warmed up again. What was I thinking? It was hot and of course the pigs wanted their water! I refilled the "pool." Waldo came, gave it a sniff, took a drink, and then walked back over to Gruffy's bucket. He gave it a push with his snout, spilling the water out onto the ground. With that, he flopped down in his do-it-yourself mud with a contented grunt. It wasn't water that he wanted, it was mud!

I walked over to his freshly filled water pool and began to shovel dirt into it to make a proper pig wallow of mud and water. Waldo couldn't wait to try it out; it was what he had wanted all along.

I've read that pigs like mud because they cannot sweat. Well, I've never seen a goat, chicken, cat, or dog sweat either, but I've seen them pant when the days are hot. So far I haven't seen either of our pigs pant, but they will make frequent visits to their mud hole on hot days. Chickens will hunch up their wings to catch a little cooler air, but the goats will sometimes lay out in the hot sun. Go figure.

No matter the human rationale for a mud hole, it is something pigs instinctively love, especially in hot weather. I would have to say I consider it a necessity for happy pigs. I know both Waldo and Polly were delighted with their new mud bath, and I'm sure Gruffy was happy to have them leave his water bucket alone.

PIG DIGS

The time had come for the pigs to have their own place. They'd been sleeping in the buck barn, but with the newly planted buck pasture starting to grow, I didn't want them rooting it up when I let the bucks back in. Plus I'd seen Gruffy and Waldo squabble over entry to the little log barn, and I didn't want that to continue. When Waldo was just a little guy, he would stand at the gate and squeal for his meal. Gruffy made great sport over picking on him. The more Waldo would squeal, the more Gruffy would push him around. It's always the smallest, youngest, and newest animals that get picked on the most and I suppose Gruffy, being the shortest goat, always got the brunt of it. Waldo was the first four-legged pasture critter that was smaller than himself. But Waldo got bigger (and heavier) and apparently didn't forget. Gruffy was now getting payback.

The more important reason for a dedicated pig shelter was to put the pigs where we needed them. Our partnership with the pigs included giving them a place to do all the rooting they wanted. As an important part of my pasture maintenance scheme they would be moved every year to the forage area that needed the most work. We wanted a shelter for the pigs that could be moved as needed.

We'd toyed with the idea of a wood structure. Many are portable: either on skids or easy to knock down and reassemble elsewhere, although our setup and equipment aren't conducive to hauling around a small structure. Another option would be to build a small pig house in each paddock. But that brought up our ongoing dilemma of time versus money. Putting up a small pig shelter sounds simple, doesn't it? But there are materials to buy and time needed for the construction. As much as Dan loves building, his time for doing so is limited because he also keeps a job off the homestead. As on all farms and homesteads, there is always more to do than time (and money) with which to do it. We try to choose our projects wisely.

After much discussion, we decided against a wood structure in favor of a temporary one—a straw bale pig house. The beauty of this is that it can be disassembled when the pigs are moved. The cattle panel and tarps can be re-used at the new location, and the straw can be used for mulch and compost right where it is. We were able to find straw from a local farm for the super-cheap price of $2 per bale. Dan bought a dozen bales and we got to work.

We chose a spot that had good drainage plus some natural shelter from the wind and rain. We started with a cattle panel, bending it into a curve and tying it to the welded-wire fence. The top, sides, and back were covered with tarps. To further secure it, Dan staked the sides of the cattle panel. We lined it with bales of straw, tying these to the panel too. There was enough tarp overhang in front to tie down for a pig height entrance. I filled it with lots of loose straw for bedding, enough for them to burrow into.

The first thing we noticed was how warm it was inside! I was very happy with that. I hoped Waldo and Polly would be happy with it too. When they finally came over to inspect it they gave it a thorough going over. I thought, "Good." But when dusk rolled around and I went to see how they were faring in their new quarters, they weren't there. Where were they?

The first place I checked was their old quarters, the buck barn. Sure enough, there they were, happily snuggled under the goat bedding. The new straw pig house was interesting, but this was home. Because I needed to get all the bucks back into the buck barn, I figured it was time to do a change-up in the routine, to encourage the pigs to use their new shelter.

Routine is the homesteader's best friend when it comes to their animals. The key is developing a routine that suits both the critters and the critter keeper. Changing a routine, however, is not all that easy. Animals don't like change; they like things they way they expect them to be.

I started by changing feeding locations. I had learned awhile back that I needed to separate the species at feeding time. The pigs and chickens will happily help themselves to the goats' feed, but as the only herbivores on the place, the goats won't touch the offerings I give to the chickens and pigs. Everybody needs to eat their own food. Once each critter learns where they are to be fed, it is fairly easy to manage keeping them all separate. The trick would be teaching both pigs and goats that the location of the chow line had changed.

There was some confusion for a while, but eventually the pigs learned that they were now fed at the gate closest to their new shelter. The bucks had to learn that they would get their feed in their paddock. At feeding time I would close the gate between the bucks and pigs, and it remained closed until the following morning. The pigs were now blocked from returning to the buck barn. With that, they chose their new pig digs, unless they happened to get to the gate before I got it shut.

While trying to establish this new routine I often found myself in a race against the pigs. It took me a few minutes after giving them their dinner before I could get to the goats with their feed. If the pigs could manage to wolf down their food quickly enough, they would race me to the gate where a double prize awaited them: the goats' feed pans plus access to the buck barn for the night.

When Dan was home we could tag team them, and eventually everyone caught on. It took several months before the pigs learned to wait by their gate for their food and not to hightail it to where the goats were eating theirs. Eventually their straw shelter felt like home, although once I opened the buck browse gate in the morning, they would often go back to visit the buck shelter to take a short nap before beginning the rest of their day.

One good thing about the straw bale shelter is that, being temporary, we can learn from it and adjust it in the future. Next time I would like to enlarge it a bit. Although it was roomy enough for two American Guinea Hogs, we discovered that they designated one corner as a manure deposit. For as much as they love mud and dirt, these animals are actually cleaner than the chickens and goats. Chickens and goats will poop anywhere and everywhere, inside their shelters or out. The pigs are free depositors outside, but inside they kept it in one spot.

All in all, the straw bale house is a good solution for a pig shelter. Perhaps we'll eventually have all-creature shelters in every paddock, but until then, it's a win-win all the way around.

Pigs in the Garden

Our garden and how we manage it has been subject to change over the years. In *5 Acres & A Dream The Book*, I told how I was starting to use permanent beds and companion plant groups. We terraced the garden for these beds instead of building raised ones, mostly because of cost. Thirty beds times the cost of lumber and soil was more than we could manage, especially with so many other needs about the place.

Several years into the permanent beds, I was exceedingly discouraged. The reason? *Cynodon dactylon*, commonly known as wire grass or devil grass. It is an uncultivated species of Bermuda grass, extremely invasive and quite prevalent in the southeastern United States. For years I battled it, but I couldn't win the war. It spreads by both seed and rhizomes, with those rhizomes growing more than 18 inches deep. It spreads over the surface of the ground in the same way. That's one reason why hand weeding is nearly impossible. That and the fact that it is as tough as wire, hence the name. Its root system will become so thick as to choke out other plants, such as my strawberries. Strawberries, I could relocate, and did so a number of times. My asparagus was another sad story, because it requires several years to establish itself before producing enough for the gardener to harvest. I finally gave up on asparagus because of the wire grass.

Heavy mulch helps, but wire grass is easily able to poke up through it and start establishing itself on top of the mulch. In the early summer I was able to stay a step or two ahead of it with continual mulching. Once harvesting and preserving commenced, however, there was no extra time for keeping up with the mulch.

Of various mulches, cardboard works best for controlling wire grass, except where cultivated plants are growing. The wire grass finds those plants' little openings in the cardboard and crowds its way up through them to start its conquest all over again.

One year I tried landscape cloth. What a mistake that was. It was expensive to buy and took days to lay down and cover with mulch. In the end the wire grass poked on up through it and, with its root system, bound the cloth to the ground. Years later, we're still dealing with shredded bits of landscape cloth.

My neighbors use Roundup weed killer and think I am nuts not to use it too. Not that I would ever use the stuff, but from surfing gardening forums, I know that this is only a temporary "solution" because no herbicide truly kills wire grass. The best poison can do is to kill the surface growth for awhile, but it always comes back because the roots are alive, well, and growing.

What's all this got to do with pigs? Am I suggesting that pigs could actually eliminate the problem? No, although I wish they could. What I was thinking, was that perhaps a different approach to the garden was in order. I knew from experience that when we used to till the garden, I had less trouble with wire grass. Oh, it eventually came back, but at least I was able to rake out quite a few of the rhizomes before I planted my seeds. That meant it took a while to reestablish itself, giving me enough time to get a harvest without having my plants choked out. And without the thick matting of roots, I could more easily plant my fall garden in the same beds.

It was that thick matting of rhizomes and roots that had me so discouraged at the end of the summer. It was time to start planting my fall garden, but after several years of becoming well established in my permanent beds, the wire grass had completely taken over. I was ready to go back to tilling.

Enter the pigs. There was a lot in the garden for which they would love to root: leftover turnips, sweet potatoes, potatoes, beets, carrots, morning glory roots, etc. Plus there were windfall remains of apples and pears. If they did as good a job in the garden as they did with the ground ivy, then I might have a pretty good chance at a decent garden the next year. This would not be a no-till scheme, but a natural till one.

As popular as no-till is with permaculturists and many gardeners, it simply wasn't working for me.

We let both the pigs and the goats into the garden once the harvest was done. I bought electric netting and a solar charger to protect the strawberries, Jerusalem artichokes, garlic, and my fall salad bed. Within about a month the pigs had beautifully turned most of the soil.

It was interesting to watch them with the electric netting. Once their snouts touched it, they steered clear. One day, however, I found Waldo on the wrong side of the netting! I realized that a recent rainstorm had knocked down a number of branches so that the electric was out. I managed to lure him back to the pig side and reestablish the charge to the netting. After that, when I let them into the garden in the morning, they would run up to the netting and stop within about a foot of it. I don't know if they could smell or feel the electricity in the net, but as long as it was on they wouldn't try to get through it.

The experiment came to an end when we began to have rubber-neckers passing by. Our road isn't especially busy, but there is a lot of local school and work traffic in the mornings and late afternoons. I reckon the sight of pigs is novel enough to slow down for a look. One day Dan went to check on the pigs in the garden and found a UPS truck stopped in the road. The driver was honking and yelling at the pigs. Dan asked if he could help him. The driver replied that he was trying to get them to lift their heads. He thought they might be wild boars and didn't think they belonged to anyone. With hunting season upon us we didn't want to take any chances of someone trespassing and killing our pigs, so they were removed from the garden.

The pigs did a fine job, however. Months later the ground they tilled remained bare, while the parts they didn't began to show the growth of cool weather weeds. From now on I am going to plan my planting to accommodate them. We probably should consider a privacy hedge along the road as well.

Polly's Piggly Wigglies

How does one know when a pig is pregnant? Hints should be that she goes into heat and that the boar tries to mate her. We saw none of this as Waldo and Polly approached breeding age, which is said to be six to eight months for males, and about eight months for females. With a gestation of three months, three weeks, and three days, a gilt can be expected to farrow for the first time around her first birthday.[1] In the absence of observed mating behavior, how does one have a clue as to whether or not piglets are to be expected?

According to Walter Jeffries of Sugar Mountain Farm, female pigs have a built-in pregnancy indicator, the lady-in-question's clitoral hood. In general, those of gilts (females that have not yet had piglets) tend to be upward pointing as the piglets they are carrying begin to weigh down her womb. The result is a shift in the connecting female anatomy.[2] This is not a strictly reliable indicator, especially as Mrs. Pig gets older, but it can be a helpful one.

In Polly's case, her piggy pregnancy pointer remained level, so I was left questioning whether or not she was in the piggy way. I would check daily, but it never changed. What did change was that her figure seemed to be filling out a bit, although that could have been from good eating. Her first birthday came and went and I was beginning to wonder if we had a dud on our hands, either her or Waldo.

My only real clue was her udder. Gradually it began to fill out, and her teats went from little bumps on her belly to fuller-looking cone shapes. Without a known due date, however, all we could do was check on her several times a day! I made a farrowing pen by attaching a hog panel to one corner of the pig shelter, and we watched and waited.

It was after lunch in early July that I made one of my Polly checks. As I approached the pig pasture I scanned the usual pig hangout spots. Waldo was busy grazing but no Polly. When I realized she was lying down in the shelter, I hollered for Dan and took off running. Polly jumped up as we approached, and there they were—six tiny newborn piglets. We had three boys and three girls.

By the time Waldo had come over for his fair share of attention we had set up the farrowing pen to surround Polly's piglets. Waldo was only mildly interested, but we decided to move him in with the bucks for the time being. We weren't certain as to how he would react to the piglets, plus I wanted to feed Polly extra and didn't want Waldo eating her food.

The question of his reaction to the piglets was answered about a week later. Within a few days they were following Polly about, including to her visits with Waldo through the gate. She and Waldo would take rests together, she on one side of the gate, he on the other. I was checking the billy boys' mineral feeder and noticed the piglets on Waldo's side of the fence. They are only about the size of a soda can at that age, and can easily slip under gates and through fences. It must have been nap time for Waldo, because he was laying on his side, completely oblivious to the three little piglets trying to figure out how to get a meal from Papa Pig!

We're definitely happy with our decision to include pigs on the homestead. They are truly remarkable animals. I'll continue to tell their tales on my blog, as we take our adventure one day at a time.

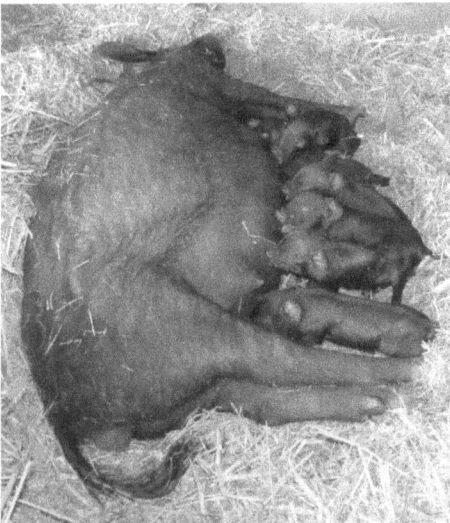

Honeybee Tales

MAKING THE COMMITMENT

Our most recent critter acquisition on the homestead has been honeybees. My interest in bees started many years ago when I lived in New Orleans, where I was working to save money for land. This was part of a group effort with friends and would eventually become my back-to-the-land experience. At the time it meant a tight budget and entertainments restricted to things that didn't cost money. A favorite was free classic movie night at Tulane University. Another was the Milton H. Latter Memorial Library, a beautiful converted mansion on St. Charles Avenue. Both of these were accessible by street car, which made them convenient for someone without other transportation. It was at the library that I stumbled upon the section on bees. I read everything it had and resolved to become a beekeeper once we were finally settled on our land.

Once I was settled, I used some of my savings to keep my promise to myself and set up one hive of Caucasian bees. These were said to be gentler than the other offering at the time—Italians. That was enough of a selling point for me. Circumstances, particularly the interpersonal workings of living with a group, did not allow me to stay long enough to harvest any honey. My one memorable experience took place during one of my first hive inspections. I was very natural at that time, and wore no bee suit, veil, or gloves. I was surprised at how heavy a full frame was. As I held up and admired that frame filled with the fruit of my honeybees' labors, it slipped from my hands and I dropped it. It crashed down on top of the open hive box and immediately produced a huge cloud of alarmed bees. I apologized profusely! To my amazement, I did not get one sting from that incident.

As we began to establish our homestead, my thoughts turned once again to bees. Our first year here, I saw not one honeybee. We had other pollinating insects, but for our first years here pollination was poor, as evidenced by our small harvests. Honeybees were a logical choice for us because during good years they produce more honey stores than their colony will use during winter. We could offer plenty of nectar and pollen sources in exchange for a little of their abundance.

Getting bees requires planning and preparation. Equipment, supplies, and bees must be ordered, and the equipment set up. Several years slipped by before I finally said, "We're just going to have to make a commitment to do this." At the end of 2014 when we discussed and wrote out our goals for the upcoming year, one hive of bees was on it. I set out to do my homework on the subject, and before we knew it, honeybees were on the way.

CHOOSING A HIVE

By now, you've probably noticed that before we add any critter to the homestead, I thoroughly research the needs and care of that critter. Granted, much of our learning is by experience, but especially in the beginning, we want to be as prepared as we possibly can be to give our animals the best life we can offer. As I researched the keeping of honeybees, I ran across the term "natural beekeeping." How could I not help but follow up on that?

Beekeeping, like any animal domestication practice, has undergone changes over the years to accommodate the keeper of the bees rather than the bees themselves. In the wild, bees will choose a cavity, such as a hollow tree, in which to establish a colony. They will begin constructing comb at the top of the cavity and work their way downward. The queen will start laying and will also move downward as the comb is filled with honey, pollen, and baby bees. After the brood hatches, the bees clean the brood cells and fill them with honey stores.[1] Once the cavity is filled, the bees will divide by swarming, with much of the colony leaving to find quarters with more room.

Modern beekeeping arranges things a bit differently. The bottom hive box is used as a brood chamber. To keep the queen, eggs, and larvae in that bottom box, a queen excluder is placed between it and the boxes ("supers") above. The queen excluder has openings sized to allow worker bees to pass through but keep the larger queen in the bottom box. With the bottom box filled with brood, the bees must store honey in the top.

To facilitate honey production, the frames in the boxes are fitted with foundation. Foundation is a sheet of beeswax or beeswax-coated plastic which is secured in the hive frames. It is embossed with the shapes of honeycomb cells to give the bees a starting point on which to build their comb. Comb is reused to save the bees' time and resources in order to make more honey.

In 1893, Professor Baudoux of Belgium proposed that larger cell sizes should produce larger bees and therefore more honey.[2] His experiments verified this and thereby paved the way for the common use of foundation imprinted with that larger cell size. Left to their own devices, however, honeybees will build comb with smaller wax cells. Even those reared in the larger cells will gradually build smaller cells over the generations.[3]

This smaller cell size is a key point in natural beekeeping. Much has been written about it and many heated discussions have ensued, especially in regard to whether or not it is a factor in controlling Varroa mites. These mites are a honeybee parasite. They lay their eggs in brood cells, the brood being the food source for the hatched mites. The mites don't kill the larvae directly, but can cause deformities and disease. Small numbers of these mites in a hive are considered normal, but too many can devastate a colony.

Proponents of the small-cell-size theory argue that problems in the honeybee industry parallel the use of larger cell sizes. They point out that Varroa mites prefer the larger drone cells in naturally-drawn comb, and that the larger-celled foundation increases Varroa counts in worker cells. It is thought that this is due not only to cell size, but also to incubation time, because large-cell workers have a longer incubation period than small-cell workers.[4] Critics of this theory point to two studies which concluded that there was no decrease in Varroa populations in small-cell hives.[5, 6]

If mites were going to be a problem either way, then there was no sense in becoming involved in the argument, nor in bothering to have an opinion about it. The appeal of the small-cell-size theory to me was that the bees—rather than the beekeeping industry—would be allowed to be the architects of their own hive. In Dan's and my way of thinking, it is vital for all of our critters to express the nature with which they were created.

I did discover one other interesting point in my honeybee research—that small-cell honeybees have a longer lifespan than large-cell bees. Typically, honeybees live about six weeks in summer, and up to six months over winter when they are at rest. The lifespan of small-cell bees is eight to twelve weeks. Why is that? Beekeepers speculate that larger honeybees wear themselves out more quickly.[7]

My conclusion from all my research was that we should let the bees live the way they would on their own, and I decided I would go with a top bar beehive. They are called "top bar" because they do not use frames with foundation. Rather, the bees are given strips of wood at the top of each box from which to build their comb. The other choice I had to make was hive style: horizontal or vertical. Horizontal top bar hives look something like a planter or bench, and have various names. Two common ones are "Kenyan" (hives with sloping sides) or "Tanzanian" (straight sides). Vertical top bar hives are commonly called "Warré" hives.

Perhaps it was because the Warré philosophy is the most hands-off and least invasive, perhaps because Warré hives are the most economical, or because they have the smallest footprint, or that they look like little pagodas...I chose to get a Warré hive. This design was developed in the first half of the 20th century by Frenchman Abbé Emile Warré. Abbé Warré believed that the keeping of bees should accommodate the bees, rather than requiring the bees to accommodate the beekeeper. His design was based on observations made during his several decades of keeping bees. He called it the People's Hive, because he thought beekeeping ought to be economical enough for everyone to do. This meant a hive design that was simple and affordable to build, even for the average person. All of this appealed to me tremendously.

Economical is somewhat relative, of course. The plans are included in Warré's book *Beekeeping For All* and can be made out of scrap lumber. The English translation of the book is available as a free download (see "Resources") or for purchase as a paperback. For those not inclined to build, beautiful ready-made hives are available, but also expensive. For my first hive I chose the middle path and bought a kit.

The hive is composed of stackable boxes which are all the same size. Two are usually used for overwintering; bees cluster in the bottom box while the top box is full of honey and pollen stores. In the spring, when plants start blooming, more boxes are added to the bottom of the hive (called "nadiring") rather than the top ("supering"). This accommodates the natural downward movement of the queen as the bees build more comb and she lays brood. As the brood hatches and leaves empty comb behind, the bees fill it with honey. All (or most) of the honey is in the top boxes, so that harvesting means simply removing the top hive box. Theoretically, it is possible to prevent swarming because the bees never run out of room! No queen excluder is required because the beekeeper knows that the queen and brood are usually in the bottom boxes. Bee escapes (one-way bee doors) can be added prior to harvest if desired, although opening the hive frequently is discouraged. The idea there is to retain the warmth and scent of the hive, which is called *Nestduftwarmebindung*.

So said the books! I was sufficiently convinced to order the hive kit and a three-pound package of honeybees. Once committed, I set about making my preparations.

ASSEMBLING AND
PREPARING THE WARRÉ HIVE

Being neither a tool person nor a builder, I asked Dan to help me assemble our hive kit. Typical Warré kits include pre-cut pieces for two to four boxes, quilt, roof, an assembled bottom board, and enough top bars for eight in each box. Mine was a four-box kit.

The hive body boxes are the most straightforward to assemble. At 300 mm by 300 mm, they are said to mimic the inside of a hollow tree. Two sides of each box have rebates, or grooves, onto which the top bars are fitted. Handles on each box complete the assembly. They are interchangeable, so there is no need to keep hive boxes in a variety of sizes.

The roof is a bit more complex. The assembly consists of a box with a gabled roof and vents: one at each of the eaves plus a ridge vent.

The "quilt" is a box with a burlap bottom. Its function is to absorb excess moisture from the hive and provide some insulation. The burlap is commonly stiffened with a flour and water paste before attaching. This is said to help prevent the bees from chewing holes through the cloth. The quilt assembly is filled with an absorbent material such as

Preassembled hive base and bottom board. Bottom boards are either solid or screened. Screen helps control Varroa mites within the colony, because when mites fall off the bees, they fall through the screen and can't climb back on.

The interior of the hive is 12 inches by 12 inches. It was designed to resemble the snug interior of a hollow log or tree.

Roof box on the completed hive with gable and ridge vents.

The quilt. This burlap-bottomed box sits on top of the hive body boxes. It is filled with an absorbant material to absorb excessive moisture in the hive, as well as provide some insulation. I filled mine with cedar wood chips.

These are the top bars, for which this style of hive is named. A thin layer of beeswax is painted along the ridge of each bar. The slots on the end of the bars fit over nails to keep them in place. A spacer bar helps determine placement.

straw, wood shavings, dried leaves, or shredded paper. I read that cedar shavings helped deter ants, so that's what I chose to use.

Bottom boards can be either solid or screened. Because of our hot summers, I chose the screened bottom to allow for better ventilation. This turned out to be a good choice for other reasons too. Most importantly, it is said to help control the Varroa mites. The mites hitch rides on honeybees and often fall off. With a solid bottom board they can climb back onto the bees and further infect the hive. A sticky paper can be placed under the screen to trap them plus keep track of their numbers. In addition, a screened bottom allows for inspection of the bees' comb-building progress in the bottom box with a mirror. This gives the beekeeper clues as to when to add more hive boxes.

Preparing the top bars required two steps. The first was to paint a thin strip of melted beeswax onto each bar's ridge. This serves as a "Start Comb Here" signal to the bees. The hoped-for result is individual combs drawn out on each top bar. Too often, the bees build comb at a diagonal, known as cross-comb. The difficulty here is that the comb is attached to two or more top bars, making it impossible to remove the bars individually for inspection or honey collection.

The second step was to place the top bars in the boxes with a 3/8-inch "bee space" between each bar. I used a 3/8-inch wide strip of wood to measure that space. Some beekeepers place the bars in the box without attaching them, but it is recommended to use brads to prevent the bars from sliding about. My top bars came with notched ends, which could slip over permanently-installed brads at the proper distance. Other Warré beekeepers nail the bars in place to prevent pulling and tearing of comb farther below when boxes are removed.

Warré beekeepers recommend keeping five boxes at the ready, but my kit came with only four. I debated what to do about this. Would I really need five boxes my first year? Should I try to build my own? Order another kit to piece out? What to do? My answer came in an email regarding my order of bees.

A CHANGE IN BEES

Chemical-free honeybees: what person dedicated to natural anything wouldn't like the sound of that? When I found an apiary in Texas which raised naturally-resistant, chemical-free bees, I was willing to pay considerably more to have a three-pound package shipped to me from across the miles. My delivery date was scheduled for April 20.

In early April I received an email from the company alerting me to a problem: UPS had suddenly changed their policy, so shipping bees with syrup was no longer permissible. Honeybees would have to be shipped with solid food. Unfortunately there was not enough time to make new shipping boxes to accommodate a dry sugar feed, so I had the choice of either canceling my order or risking the U.S. Postal Service (known to be slower with such deliveries). I opted to receive a refund, but it left me frantically trying to find bees.

Internet searches for chemical-free bees yielded limited results. The apiaries that I did find had long since been sold out. I finally contacted a local beekeeping supplier and learned that they did indeed still have packages available. I was told they were "hygienic" bees, because they had been bred to monitor the brood comb and remove dead, diseased,

My refund allowed me to purchase this observation hive box.

or infested larvae and pupae. They did not come from a strictly chemical-free apiary, but they were considerably cheaper than my Texas bees and I could pick them up myself rather than worrying about UPS or the U.S. Postal Service.

I decided to invest the rest of my refund in more beekeeping equipment. I made two purchases: a Warré observation hive box and a top feeder. The observation box came with a Plexiglas window which could be uncovered to take a peek into the hive. This gave me the recommended five body boxes, so I felt better prepared. The top feeder fit exactly on the top hive box under the quilt and held close to a gallon of sugar syrup. The advantage was that it could feed hundreds of bees simultaneously, rather than only the several dozen of a typical inverted-jar feeder.

My new bee-arrival date was April 18th. I wrote out my last-minute to-do list with no little excitement. This was an important addition to the homestead, one that could finally be crossed off our list of self-sufficiency goals.

Essential Oils for Honeybees

If natural beekeeping doesn't use chemicals, how does it deal with bee pests and diseases, especially Varroa? Many Warré beekeepers are strictly hands-off in this matter. They rely on screened-bottom boards, smaller natural cell-size, and the bees themselves to keep Varroa numbers to a minimum. Bee losses from Varroa are considered the lesser of two evils versus using chemicals. Others use what are considered natural substances for Varroa mite control. I've already discussed the screened-bottom board and small cell-size, but what about some of these other practices?

Bee behaviors can affect Varroa populations. Some bees take a more active role in monitoring and removing infected larvae. Apparently, this is a genetic trait.[1] The bees I was getting were said to be hygienic, but was there anything else I could do to minimize problems?

Researchers at West Virginia University experimented with essential oils (EOs) and discovered that certain ones—wintergreen and spearmint—are toxic to Varroa mites. The easiest application was mixing both EOs with sugar syrup for the bees to consume. However, the bees did not particularly care for mint. The other problem was that the oils floated to the top of the syrup. Once the researchers added lemongrass essential oil and lecithin as an emulsifying agent, the bees readily and effectively took the syrup.[2] To a 1:1 (one part water and one part sugar) syrup, add half an ounce each of the essential oils and a capful of liquid lecithin to make up a quart. This is mixed in a blender and added to the bees' sugar or honey syrup at one tablespoon per gallon.[3]

Besides helping to control Varroa mites, other claims for the use of these oils include deterring absconding (which is where the bees, for whatever reason, decide they don't like their new hive and simply leave), encouraging faster comb-building, helping to prevent rejection of a new queen, and helping to calm the bees. In addition, lemongrass essential oil, which contains some of the same natural pheromones that queen bees use to attract workers, has been shown to reduce the bees' urges to sting.[4] The only negative is that it promotes robbing because the bees love it so much.

I was starting with packaged bees which meant that my bees should be relatively Varroa-free. But I liked the other benefits of the syrup, so I bought the essential oils and made it. I wanted to start my natural beekeeping off on the right foot.

Bee Day

It had been raining almost daily for weeks. As honeybee-pickup day approached, there was concern around the county as to whether there would be a postponement or not. The bee people sent out an email several days prior, assuring us our bees would be ready. Even so, I wondered about trying to hive them in the rain. The scheduled Saturday dawned sunny and blue-skied, much to every beekeeper's relief, me included.

I originally planned to bring the package inside on the back porch temporarily, but changed that to the bench outside the porch door. That was because the bee pick-up area had bees flying everywhere and the little screen-sided boxes had dozens of loose bees on them, including the one the gentleman handed to me. All the way home I could hear buzzing in the back of my Jeep, and a few loose bees were flying around the car. After I got them unloaded, I lightly sprayed the package screens with sugar syrup mixed with the essential oils. By the time I came back out with my camera, all was quiet as the bees feasted.

The basic procedure for hiving bees is to first remove the can of sugar syrup, remove the little cage containing the queen, place the queen cage in the hive, and then dump the bees out of the shipping container on top of her. As daunting as this sounds, honeybees are not inclined to sting at this point, so the chances of the beekeeper getting attacked are relatively small. By nature, bees are defensive, not aggressive, and newly-packaged bees have nothing to defend. They have no home, stores, nor brood; even the queen is not familiar to them. Because of all that, the smoker is not necessary. I had watched quite a few videos on this subject, with quite a few brave souls wearing only veils for this process. I was not quite that brave so I had donned my bee jacket and veil, because I knew there would be a lot of bees flying around.

I set about in the calmest manner I could muster to introduce my bees to their new home. I kept them lightly sprayed with sugar syrup at intervals as I set up my work area. I also sprayed the inside of my hive in hopes of preventing absconding. I pried off the cover of their box, found the syrup can, and slowly pulled the can out, making sure to turn it upright when I set it next to me on the ground. The next step was to find the packing strap stapled near the opening and pull out the queen cage. I slowly and carefully pulled on the strap to discover...no queen cage! Now what? I was pretty sure it was on the bottom of the package under a pile of bees. I quickly called the bee people who told me to dump the bees into the hive and try to catch the queen cage as it

I had quite a few bees left in the shipping box after dumping the contents into the hive. I placed it near the hive entrance for stragglers to find their way.

fell out. Just what I wanted to hear, especially since I hadn't yet purchased bee gloves! I had procrastinated getting those gloves because I hadn't planned to need them for this procedure. I knew gloves would be handy for certain situations; I just wasn't expecting this to be one of them. I grabbed Dan's welding gloves which were too big and therefore awkward, but I managed to catch the queen cage and proceed as I was supposed to.

My initial hive set-up included two hive body boxes. The top bars were removed from the upper box in order to dump the bees into the hive, while the bottom box had its top bars in place. It is onto these that the queen cage is placed before shaking the bees out on top of it. The queen cage has two holes, one plugged with sugar candy and a cork, the other with only a cork. The cork is removed from the candy end and a small hole is carefully poked through the soft candy with a long nail. This gives the bees a start for freeing their new queen. By the time they have eaten their way through the candy, they will be accustomed to her scent and readily accept her as the official mother of their hive.

Once the queen cage was properly placed and the bees dumped into the hive, I replaced the top bars in the top box, added the feeder, the quilt, and lastly the roof. I set the almost empty bee shipping package next to the hive opening, so that any remaining bees could hopefully find their way into the hive. I worried about them a bit because I hadn't managed to dump out as many bees as I had hoped.

The next morning it rained again. I was glad to see that the package was empty, with a few bees flying around and on the outside of the hive. I checked their feeder and added more syrup. Lots of bees were busy feeding, which made me happy. In a few days I could check to make sure the queen had been released. In the meantime, the only thing left to do was to let them be.

FIRST HIVE CHECK

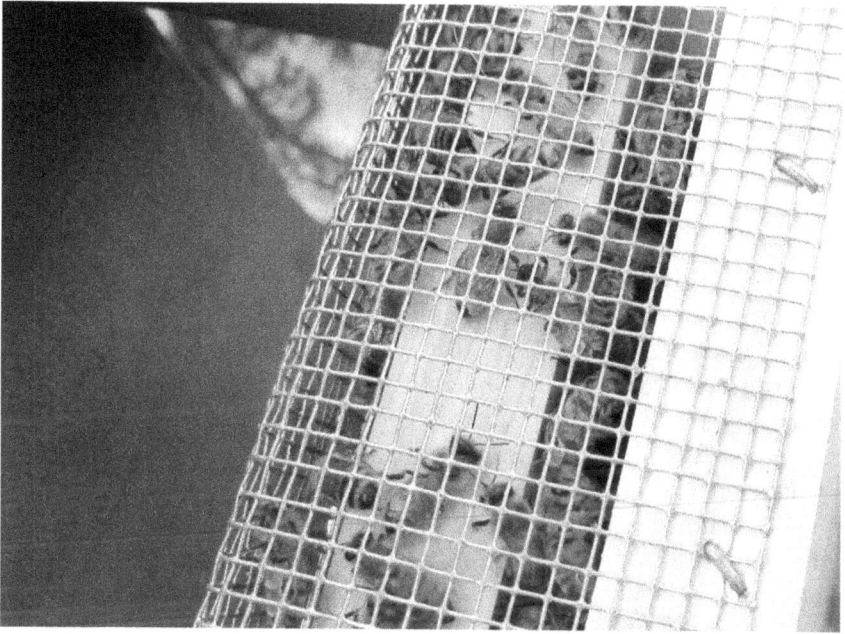

We had deluge forecasted for the remainder of the week, so on the first clear day I donned my bee suit and lit my smoker to make my first official check on my honeybee hive. The goal was to check the queen cage to see if the queen had been released.

I directed a few puffs of smoke into the hive entrance and waited a bit. Then I removed the roof and quilt. Many bees were busy at the feeder and seemed to take no notice. My Warré top feeder fits squarely on the top hive box. Screening enables the bees to enter the feeder and feast on the sugar syrup with added essential oils.

Next I removed the feeder. I was amazed at how much comb had already been built. It was snowy white and appeared to be following the top bars along the wax beads I had painted on the undersides of the bars. I was happy to see that. Next, I removed the top hive box and gently set it aside, comb, bees, and all. The bees appeared not to notice.

I found the queen cage where I'd placed it, but under a pile of bees. I poofed the bees away with my bee smoker and discovered that the queen was still in her cage. It had only been a couple of days so I could have left her in longer, but considering the weather I decided to release her. Opinions vary on how long to give the bees to get her out, but most agree that if she's been with them for at least two days she will usually be accepted.

I removed the cork, set the cage back down, and she walked right out, heading immediately into the bottom hive box and away from the light. I quickly picked up the now empty queen cage and reassembled the hive. Before putting the roof back on, I topped off the feeder and left the bees to get on with their bee business. As soon as I got back into the house it began to pour!

My sense of relief for having accomplished that task was quickly replaced with worry. Should I have given the bees more time to release their queen? Had I taken matters into my own hands prematurely? What if they didn't accept her? What if they killed her?

Apparently I wasn't the only one with such questions, because it was a topic of discussion on an online beekeeping group I had joined, called "warrebeekeeping" (see "Resources"). Experienced members of the group said to watch for pollen being brought in. Pollen is the bees' source of protein and is mixed with nectar and bee secretions in what is known as "bee bread." It is fed to baby bees (larvae). If the bees are bringing in a good amount, it means the queen is present and laying.

I had no idea what a "good amount" was, so all I could do was watch and wait. There was only a trickle of pollen at first, but I noticed that pollen deliveries increased as the days passed. I took that as a good sign. Still, I remained anxious.

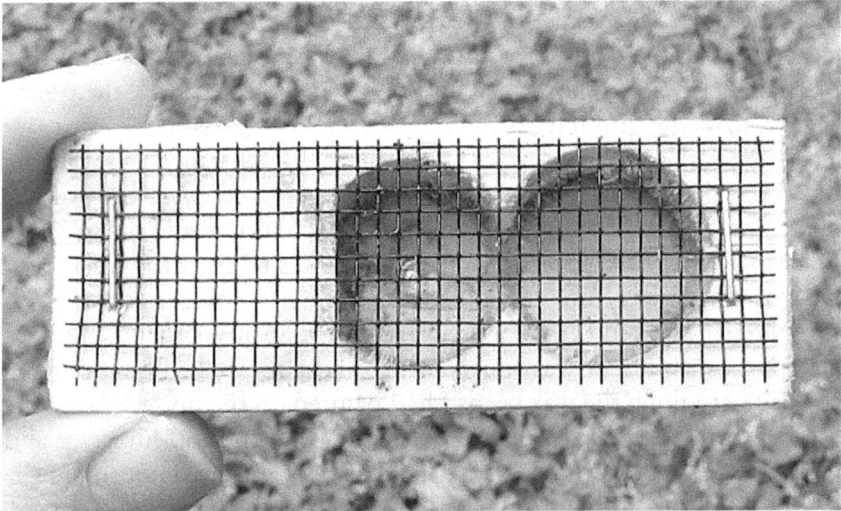

The empty queen cage. It is designed to accomodate one queen and several attendants. The chamber on the left is filled with soft bee candy to feed the queen. The worker bees eat through the candy to free the queen from her cage.

SECOND HIVE CHECK

Two of my questions as a novice Warré beekeeper were: how long should I feed my bees after I placed them in my hive, and when would I need to add more hive boxes? Opinions on feeding seem to vary greatly: from "not at all," to "until they fill the entire top box with comb," to "until they stop consuming the sugar syrup." When our new colony was about a month old, I made my second hive check to see how far along they were in building comb in the top hive box. Dan and I lit the bee smoker, removed the top, quilt, and feeder, and took a look.

The first thing I noticed was that the top box was becoming quite full of comb, and that all of it appeared to be built along the length of the top bars. I also noticed that some of the comb was already capped and that several of the top bars still had no comb.

I did not remove any of the top bars to look for the queen or brood. Pollen had been coming in regularly and between that and the bees' behavior, I trusted that my hive was queenright (had a healthy queen). The bees were busy with the comb and seemed to be acting with purpose—all good signs of a happy, healthy hive. They had emptied the feeder, so I added more syrup. I figured that would be the last time I would have to do it.

This inspection also helped answer my second question: how was I to know when to add another hive box? The idea is to enlarge the hive before they fill the available space and decide to swarm in search of a larger home. According to David Heaf in his *Natural Beekeeping with the Warré Hive*, boxes are nadired (added to the bottom) when the bottom box is about half-built with comb.[1] I started with two boxes, and since my bees were still working in the top box, I knew it wasn't time yet.

Satisfied with what we saw, we carefully put the roof back on and left our busy bees to carry on with their work.

ENLARGING THE HIVE

Seated on an overturned bucket, I watch bees coming and going. This has become part of my daily routine, not a chore, but a pleasant little break from other things which must be done around the homestead. My bees are fascinating to watch, because I know they do all things with a purpose. I see pollen coming in regularly and have learned that there are different colors of pollen, ranging from various shades of yellow and gold to green.

I know that the bees I'm seeing are the oldest, because a bee's activity is largely determined by her age and physical development.[1] I say "her," because most of the bees are worker bees, which are female. Newly emerged workers begin their work in the brood area. They start with cleaning out cells, and then feeding larvae and the queen. Soon the young worker is receiving nectar from returning foragers. Older house bees also have waste removal duty.

At two to three weeks of age a worker bee's wings are developed and she takes on guard bee duties. Soon she becomes a scout or forager, either scouting out new pollen and nectar sources, or collecting and bringing them back to the hive. Bees also forage for water, plus plant resins which they use to make propolis. Propolis is the sticky bee glue which is used to seal the interior of the hive against drafts and rain.

Checking comb-building progress in the bottom hive box became a weekly chore after our second hive check. If the bees begin to run out of room, they will swarm. They will build queen cells to leave the remaining bees a new queen, but the bulk of the colony will go off in search of more spacious quarters. This is actually one way beekeepers add hives: by catching swarms. As much as I wanted more hives, however, I wasn't ready for that!

Is there a way to check comb-building progress without opening up the hive? The answer to that is, "Yes." One way is to build observation windows into the hive boxes. I had one box with such a window but did not use it when I first set up the hive. Another way is to use a handheld mirror. That's where a screened bottom comes in handy, because I can use the mirror to look up through the bottom of the hive. I took my first peek about a week-and-a-half into my beekeeping, but I didn't see any bees reflected in the mirror. I knew this was because they were busy working in the upper box of the hive. As they filled that box with comb, I began to see bees in my mirror near the top bars of that bottom box.

As our beekeeping approached its two-month anniversary, I could see that the bees were busy working on comb-building in that bottom box. I couldn't tell if it was half-full yet, but the last thing I wanted was for my bees to swarm. We chose a sunny Saturday to add more boxes to expand the hive. Unlike conventional beekeeping, however, we would not be adding boxes to the top (supering); we would be adding boxes to the bottom (nadiring).

For us, nadiring was definitely a two-person job requiring one person (the stronger of the two – okay, Dan) to lift the hive to allow a second (me) to place empty boxes underneath. Some folks use hive lifts, but as novice beekeepers, we only had the simplest equipment. Some Warré beekeepers add enough boxes in the spring to accommodate the year's honey flow. Others warn that this makes the hive top-heavy, which could be disastrous in areas of high winds. Nadiring one box at a time, or at least strapping down the hive, is therefore recommended.

I decided we would add two more empties to the hive. I did not know if this was the correct decision, but without a hive lift, it seemed better—at least for our backs—to hoist only one full box than two. I had chosen to locate the hive against a fence, so I wasn't too concerned about it toppling over in high winds. The fence would also protect the hive from spying eyes. Believe it or not, beehive theft is a problem in some areas.

Lifting full, heavy hive boxes can require a strong back.
Alternatively, a hive lift can be made to do the lifting.

We suited up and lit the smoker. Lots of bees were busy coming and going. I puffed a little smoke into the hive entrance, removed the roof, and set it aside. We did not remove the quilt, but I did give the shavings a stir to check for moisture. We puffed some more smoke into the hive and then used the hive tool to gently pry the bottom box from the stand. I was surprised at the amount of propolis I saw; I thought there would have been more based on the amount of time that had passed. Next, Dan hoisted the two boxes and I quickly placed two more on the stand. We reassembled the hive and were done. And the bees? Amazingly, unimpressed. In fact, they appeared not to notice. Activity continued as though nothing had happened.

BEES IN OUR FUTURE

Although we have yet to harvest our first drop of honey, I am already looking forward to adding more hives to the homestead. Even without the honey, we need the bees for pollination. We've planted quite a few fruit and nut trees and bushes: apples, pears, peaches, elderberries, plums, cherries, mulberries, blueberries, raspberries, persimmons, almonds, and hazelnuts. Then there are the vegetable and herb gardens, plus field crops like cowpeas. Everything would benefit from better pollination.

One piece of equipment I have not had to buy is a honey extractor. This drum-like apparatus contains racks which hold the frames in place and uses centrifugal force to extract the honey. Even though it can be used for top bar comb, I am not interested in using it. The simplest method for harvesting honey is to crush the comb and let it drain. Some beekeepers believe this method is inefficient, because it means the bees must continually draw out new comb. They observe that bees that can reuse comb can produce brood and honey more quickly, resulting in a larger harvest for the beekeeper. Others say the bees will actually draw out their own comb faster with top bars rather than on foundation, because the bees aren't keen on plastic. However, I want to use the beeswax in other ways on my homestead, so I am willing to ask my bees to spend time making both comb and honey rather than just honey.

Bees have four pairs of glands in their abdomens which secrete scales of wax. They use these scales to make comb. I look forward to using the beeswax to make candles, salves, and lotions. In pondering a sustainable emergency source of lighting, beeswax candles seem to be the most feasible for us. We have a few kerosene lamps for emergency use, but kerosene is something I cannot produce on the homestead. Also, I must replace it if it sits too long, because kerosene has a shelf life. My rationale for making my own salves and lotions is similar; it is one more thing I can do for myself, with the additional assurance that the ingredients are simple and pure.

I think another benefit to having the bees draw out their own comb is that it facilitates the removal of old and dirty comb more quickly than traditional beekeeping often allows. Reusing comb is understandable, because foundation costs money and takes time to install. On the other hand, it can harbor hidden pests and disease. Top bar beekeeping eliminates those concerns because the old comb is removed with each harvest.

While I await my first harvests of honey and beeswax, I focus on planting things that will feed my honeybees. Technically called "nectary plants," these provide pollen and nectar for insects, including bees. Types of nectary plants vary from region to region (see "Resources" for a helpful website). Nectary plants appropriate for the southeastern US are herbs such as angelica, anise hyssop, basil, borage, chives, lemon balm, various mints, oregano, rosemary, sage, and thyme; field plants and cover crops such as buckwheat, clover, cowpeas, and sunflowers; garden plants such as squashes, melons, and cucumbers; plus all manner of fruits and nuts. My goal is to have something blooming every possible month of the year.

Next spring I plan to add two more Warré hives to our apiary. After that, who knows? I don't necessarily have a set number of hives in mind; I just love having honeybees around. My first hive was a kit, but I may even be inspired to learn the necessary woodworking skills to build my own. That would be something for me.

As with our other homestead critters, I will continue to read, study, and tackle problems as they present themselves. Next time around, I hope to have many more honeybee tales to tell.

POSTSCRIPT

Critter Tales is the kind of book that is difficult to stop writing. The antics of my critters are ongoing, the stories never-ending. Every day there is some new incident, something amusing, something learned, something new to tell. But eventually, it is time to stop writing and get on with the business of editing, formatting, and designing the book both inside and out. This postscript serves as a last minute update on the critters to which I have introduced you.

Of our honeybees there is little more to tell, except that I bought two more hive kits. Because of our late start with beekeeping, we will let them keep all their honey stores for the winter. To their credit, the bees are the least demanding critters on the homestead.

We continue to be dogless, but we have acquired another cat. My daughter-in-law's young cat was not happy as an indoor-only kitty, so she came to live with us. She is an amazing hunter and has proved to be the best farm cat ever!

Of chickens, my Buff Orpingtons have become older ladies which no longer give me many eggs. We are gradually preparing them for chicken stew, while the up-and-comers are Black Australorps. Of these, we have a dozen pullets and two roosters, so expect more rooster tales in the near future.

Goat breeding season is upon us once again, and my little Kinder herd is growing. At the moment I have six does, two bucks, and one of last year's bucklings to provide us with chevon. That's quite a few goats, and I continue to analyze what is the best number to keep a balance with our land.

Of Polly's six piglets, we always planned on keeping two for bacon and sausage, and that's what we have left. I sold three gilts. Of the remaining piglet, I made a nice trade. I traded one little piglet for six...

Muscovy Ducks

ENDNOTES

Chicken Tales

DOING MY HOMEWORK

1 Leigh Tate, "Unofficial Egg Laying Research Results," *5 Acres & A Dream The Blog*, last modified May 9, 2011, accessed September 29, 2015, http://www.5acresandadream.com/2011/05/unofficial-egg-laying-research-results.html.

2 Heather Harris, "How To Tell If A Chicken Is Still Laying," *The Homesteading Hippy,* last modified November 25, 2013, accessed July 13, 2014, http://thehomesteadinghippy.com/how-to-tell-if-a-chicken-is-still-laying/.

EGG!

1 Cheryl Long and Tabitha Alterman, "Meet Real Free-Range Eggs," *Mother Earth News*, October/November 2007, accessed September 29, 2015, http://www.motherearthnews.com/real-food/free-range-eggs-zmaz07onzgoe.aspx.

2 Tabitha Alterman, "Eggciting News!!!," *Mother Earth News*, last modified October 15, 2008, accessed September 29, 2015, http://www.motherearthnews.com/real-food/pastured-eggs-vitamin-d-content.aspx.

3 Long and Alterman, "Meet Real Free-Range Eggs."

BROODY

1 "Welsummer," *Wikipedia*, accessed July 17, 2014, https://en.wikipedia.org/wiki/Welsummer.

2 Carla Emery, *The Encyclopedia of Country Living: An Old Fashioned Recipe Book*. 9th ed. (Seattle: Sasquatch Books, 1998).

3 ibid.

4 "Broodiness," *Wikipedia,* accessed June 30, 2014, https://en.wikipedia.org/wiki/Broodiness.

5 *5 Acres & A Dream The Blog*, http://www.5acresandadream.com.

MRS. MEAN

1 "Holland Chicken," *The Livestock Conservancy*, accessed July 17, 2014. http://www.livestockconservancy.org/index.php/heritage/internal/holland.

MYSTERY OF THE DISAPPEARING CHICKS

1 Elwood Lightfoot, "Best way I've found yet to deal with snake problems!!," *Backyard Chickens*, last modified June 1, 2011, accessed June 12, 2013, http://www.backyardchickens.com/t/515899/best-way-ive-found-yet-to-deal-with-snake-problems.

2 Handyman, "Homemade rat trap," last modified December 3, 2012, accessed June 12, 2013, https://www.youtube.com/watch?v=OXRgZLs6OJo.

MY PERSONAL CHICKEN

1 "Ameraucana Breed Standards," *Ameraucana Breeders Club*, accessed September 30, 2015, http://ameraucana.org/standard.html.

MOVING DAY FOR CHICKENS

1 Carol Ekarius, *How To Build Animal Housing* (N.p.: Storey Publishing, 2004), 59.

2 "Backyard Poultry and Pigeon Houses," Penn State Cooperative Extension Idea Plan No. IP 727-25.

3 Lydia Ray Balderston, *Housewifery: A Manual and Text Book of Practical Housekeeping*, 1921 (N.p.: Forgotten Books, 2012), 300.

OF CHICKENS, GOALS, AND WHAT I'VE LEARNED

1 Leigh Tate, "Appendix B, "*5 Acres & A Dream The Book* (N.p.: Kikobian Books, 2013).

2 Geoff Lawton, "How to Grow Chickens Without Buying Them Any Grain By Only Feeding Them Compost," *Geoff Lawton*, accessed April 17, 2015, http://www.geofflawton.com/.

3 Katherine Grossman, "How to Preserve Eggs With Water Glass," *Granny Miller: A Journal Of Agrarian Life & Skills*, accessed April 17, 2015, http://www.granny-miller.com/how-to-preserve-eggs-with-water-glass/.

4 "Herbert Hoover," *PresidentsUSA*, accessed April 17, 2015, http://www.presidentsusa.net/1928slogan.html.

5 J. I. Rodale and Staff, ed., *Encyclopedia of Organic Gardening* (Emmaus, PA: Rodale Books Inc, 1999), 689.

Goat Tales

BUT FIRST, FENCING

1 Leigh Tate, *5 Acres & A Dream The Book* (N.p.:Kikobian Books, 2013), 213.

Jasmine

1 Dr. Susan Kerr, "Drying Off Lactating Livestock," *Oregon State University Small Farms,* Summer 2010, accessed August 16, 2014, http://smallfarms.oregonstate.edu/sfn/su10dryinglivestock.

2 "Garlic," *Phytochemicals,* accessed August 17, 2014, http://www.phytochemicals.info/plants/garlic.php.

3 Dr. John R. Christopher, "Bone Flesh, & Cartilage," *Dr. Christopher's Newsletter 1-8,* accessed August 16, 2014, http://www.herballegacy.com/ Bone_Flesh_&_Cartilage.html.

4 Mark Steel, "Herbs, Milking and Mastitis, Five Secrets," *Back2theLand,* last modified August 2, 2009, accessed August 17, 2014, http://back2theland.com/health/herbs-milking-and-mastitis-five-secrets/.

5 Pat Coleby, *Natural Goat Care* (Austin, TX: Acres U.S.A., 2001), 241.

The Game Changers

1 "Goat Breed Comparison Chart," *The Livestock Conservancy,* accessed August 21, 2014, http://albc-usa.etapwss.com/index.php/ heritage/internal/goat-chart.

2 "Arapawa Goat," *Wikipedia,* accessed August 21, 2014, https://en.wikipedia.org/wiki/Arapawa_Goat.

It Was Time For Elvis To Go

1 Lierre Keith, *The Vegetarian Myth* (Crescent City, CA: Flashpoint Press, 2009), 5.

2 Joel Salatin, *You Can Farm* (Swoope, VA: Polyface, Inc., 1998), 198.

3 Scott and Helen Nearing, *Living The Good Life* (New York: Schoken Books, Inc., 1970), and *Continuing The Good Life* (New York: Schoken Books Inc., 1979).

4 Andrew Schreiber, "On Killing and Eating Animals: thoughts on the role of animals in creating & sustaining human communities," *Windward Education and Research Center,* last modified September 28, 2013, accessed April 14, 2015, http://www.windward.org/2.0/notes/2013/2013andrew15.htm.

Goat Midwifery: Learning the Hard Way

1 Molly Nolte, "Ketosis and Pregnancy Toxemia," *Fias Co Farm,* last modified August 15, 2012, accessed August 20, 2014, http://fiascofarm.com/goats/ketosis.htm.

2 ibid.

3 Coleby, *Natural Goat Care,* 154.

4 "What is Chaffhaye?," *Chaffhaye,* last modified 2012, accessed October 13, 2014, http://www.chaffhaye.com/what-is-chaffhaye/.

5 Sue Reith, "Hypocalcemia - Ca and Ph in the diet," *Dairy Goat Care and Management*, last modified January 7, 2007, accessed October 13, 2014, http://goats.wikifoundry.com/page/Hypocalcemia+-+Ca+and+Ph+in+the+diet.

SOLVED: MYSTERY OF THE DYING KIDS

1 Jack and Anita Mauldin, "Symptoms," *Mauldin Boer Goats*, accessed September 9, 2014, <http://www.jackmauldin.com/symptoms.html>.
2 Pat Showalter, "Re: 3 or 4 babies-enough milk to share?," *Kinder Goat Yahoo Group*, Message 16875, July 22, 2014, https://groups.yahoo.com/neo/groups/KinderGoats/conversations/messages/16875.
3 Tate, *5 Acres & A Dream The Book*, 220.

TOWARD SUSTAINABLE GOAT KEEPING

1 Suzanne Gasparotto, "Tall Fescue Toxicity in Goats," *Onion Creek Ranch*, accessed November 24, 2014, http://www.tennesseemeatgoats.com/articles2/fescue.html.
2 Dave Jacke, *Edible Forest Gardens Volume 2* (White River Junction, VT: Chelsea Green Publishing, 2005), 616.
3 Susan Schoenian, "The truth about grain: Feeding grain to small ruminants," *Maryland Small Ruminant* Page, last modified 2007, accessed November 24, 2014, http://www.sheepandgoat.com/#!truthgrain/cjjy.
4 Paul Dettloff, DVM, *Alternative Treatments for Ruminant Animals* (Austin, TX: Acres U.S.A., 2009), 9-12.
5 Schoenian, "The truth about grain."
6 F. B. Morrison, *Feeds and Feeding: A Handbook for the Student and Stockman* (Ithaca NY: The Morrison Publishing Co., 1943) 27-28.
7 Detloff, *Alternative Treatments for Ruminant Animals*, 9-12 .
8 Suzanne Gasparotto, "Long Fiber: Critical to Good Nutrition," *Onion Creek Ranch*, accessed November 24, 2014, http://www.tennesseemeatgoats.com/articles2/longfiber06.html.
9 M. Hadjipanayiotou, "Effect of Sodium Bicarbonate and of Roughage on Milk Yield and Milk Composition of Goats and on Rumen Fermentation of Sheep," *Journal of Dairy Science* 65, no. 1 (January 1982), accessed November 24, 2014, http://www.sciencedirect.com/science/article/pii/S002203028282153X.
10 Gasparotto, "Long Fiber: Critical to Good Nutrition."
11 R. Bowen, "Nutrient Absorption and Utilization in Ruminants," *Colorado State University*, last modified 2009, accessed December

12, 2014, http://www.vivo.colostate.edu/hbooks/pathphys/ digestion/herbivores/rum_absorb.html.

12 Robert J. Van Saun, DVM, "Feeding For Two," *Cornell University*, accessed November 24, 2014, http://www.ansci.cornell.edu/goats/ Resources/GoatArticles/GoatFeeding/FeedingForTwo.pdf.

13 Robert L. Johnson, "The Feeding Of Goats," *International Dairy Goat Registry*, last modified April 29, 1996, accessed November 24, 2014, http://idgr.info/index/articles/the-feeding-of-goats.

14 Keenan Bishop, "Getting your goat and feeding it too," *The State Journal*, last modified October 7, 2007, accessed November 26, 2014, http://www.state-journal.com/spectrum/2007/10/07/getting-your-goat-and-feeding-it-too.

15 Coleby, *Natural Goat Care*, 95 - 97.

16 Gunnar Sundstøl Eriksen, "Effects of phyto- and mycoestrogens in domestic animals," accessed November 26, 2014, http://www.dnva. no/binfil/download.php?tid=48848.

17 Carol Raczykowski, "Hormonal Causes of Infertility in the Doe," *Pygmy Goat WORLD*, last modified 1994, accessed November 26, 2014, http://kinne.net/infert1.htm.

18 Coleby, *Natural Goat Care*, 95 - 97.

19 Kristie Miller, "Goat Feeding Regime," *Land of Havilah Farm*, accessed November 26, 2014, http://www.landofhavilahfarm. com/loh-feed-regimen.htm.

20 "Nutritional Requirements of Goats," *The Merck Veterinary Manual*, accessed November 26, 2014, http://www.merckmanuals. com/vet/management_and_nutrition/nutrition_goats/nutritional _requirements_of_goats.html.

21 Johnson," The Feeding Of Goats."

22 "Crude Protein: Ruminant Nutrition, CNCPS, Crude Protein Fractions, Rumen (un)degradable protein," *DietaryFiberFood.com*, last modified, April 5, 2012, accessed January 19, 2015, http://www. dietaryfiberfood.com/protein/crude-protein-from-animal-nutrtion-perspective.php.

23 Van Saun, "Feeding For Two."

Llama Tales

A Worrisome Turn of Events

1 Mark D. Leichty and Ila A. Davis, "Llama Failure to Thrive Syndrome," *Iowa State University Veterinarian* 54, no. 2 (1992), accessed October 1, 2015, http://lib.dr.iastate.edu/iowastate _veterinarian/vol54/iss2/9.

Puppy Tales

POOR KRIS

1 "OT: puppy with elbow dysplasia," *Holistic-Goats*, Message 56424, accessed May 10, 2012, https://groups.yahoo.com/neo/groups/Holistic-Goats.

2 Ray Peat, "Gelatin, stress, longevity," *RayPeat.com*, last modified 2009, accessed January 20, 2015, http://raypeat.com/articles/articles/gelatin.

Guinea Tales

TO TAME OR NOT TO TAME?

1 Jeannette S. Ferguson, "How Tame Are Your Guineas?," *Frit's Farm*, accessed October 4, 2015, http://www.guineafowl.com/fritsfarm/guineas/tamed/.

Kitty Tales

POISONED

1 Jean Dodds, DVM, "Garlic: Beneficial or Harmful to Companion Animals?," *Pet Health Resource Blog*, last modified January 17, 2013, accessed March 3, 2015, http://drjeandoddspethealthresource.tumblr.com/post/40797875107/garlic-beneficial-harmful-to-cats-dogs.

Pig Tales

CONSIDERING PIGS

1 "Red Wattle Hog," *The Livestock Conservancy*, accessed January 12, 2015, <http://www.livestockconservancy.org/index.php/heritage/internal/redwattle>.

2 "American Guinea Hog," *The Livestock Conservancy*, accessed January 12, 2015, http://www.livestockconservancy.org/index.php/heritage/internal/guineahog.

3 *American Guinea Hog Association*, accessed January 12, 2015, http://guineahogs.org/articles-about-guinea-hogs.

4 "Raising American Guinea Hogs: Fencing, Housing, Food and Water," *Quartz Ridge Ranch*, last modified December 6, 2012, accessed January 12, 2015, http://quartzridgeranch.wordpress.com/2012/12/06/raising-american-guinea-hogs-fencing-housing-food-and-water/.

POLLY'S PIGGLY WIGGLIES

1 "Life Cycle of American Guinea Hogs," *American Guinea Hog Association*, accessed July 17, 2015, http://guineahogs.org/life-cycle-of-american-guinea-hogs/.

2 Walter Jeffries, "Pregnancy Indicator," *Sugar Mountain Farm*, last modified August 28, 2011, accessed July 17, 2015, http://sugarmtn farm.com/2011/08/28/pregnancy-indicator/.

Honeybee Tales

CHOOSING A HIVE

1 Dustin Bajer, "Natural Beekeeping with Warre Top Bar Hives," last modified February 8, 2014, accessed May 15, 2015, http://dustin bajer.com/natural-beekeeping-warre-top-bar-hives/.

2 Dee A. Lusby, "Honeybee Comb: Brief History, Size and Ramifications – Part 1," accessed May 13, 2015, http://www. beesource.com/point-of-view/ed-dee-lusby/honeybee-comb-brief-history-size-and-ramifications-part-1/.

3 Sam Comfort, "Less Invasive Beekeeping," *Anarchy Apiaries*, accessed May 10, 2015, http://anarchyapiaries.org/hivetools/node/32.

4 "Cell Size," *Resistantbees.com*, accessed June 18, 2015, http://resistantbees.com/zelle_e.html.

5 Jennifer A. Berry, William B. Owens, and Keith S. Delaplane, "Small-cell comb foundation does not impede Varroa mite population growth in honey bee colonies," *Apidologie* 41 (March 20, 2008): 40-44, accessed June 19, 2015, http://www.ent.uga.edu/bees/documents/m08138.pdf.

6 Thomas D. Seeley and Sean R. Griffin, "Small-cell comb does not control Varroa mites in colonies of honeybees of European origin," *Apidologie* 42 (2010): 526-532, accessed June 10, 2015, http://www.reeis.usda.gov/web/crisprojectpages/0211868-evaluation-of-small-cell-combs-for-control-of-varroa-mites-in-new-york-honey-bees.html.

7 "Hygienic behavior," *Resistantbees.com*, accessed June 18, 2015, http://resistantbees.com/hyg_e.html.

ESSENTIAL OILS FOR HONEYBEES

1 John Harbro, "VSH-Based Resistance to Varroa," *Harbro Bee Co.*, accessed May 20, 2015, http://www.harbobeeco.com/vsh/.

2 Jim Amrine, Bob Noel, Harry Mallow, Terry Stasny, and Robert Skidmore, "Results of Research: Using Essential Oils for Honey

Bee Mite Control," *WVU*, last updated December 30, 1996, accessed May 22, 2015, http://www.wvu.edu/~agexten/varroa/varroa2.htm.

3 "Honey Bee Healthy recipe," *The Wasatch Beekeepers Association*, accessed May 22, 2015, http://www.wasatchbeekeepers.com/443-2/.

4 Bob Noel, "Honey-B-Healthy," accessed May 22, 2015, http://www.rnoel.50megs.com/john/index.html.

RESOURCES

Here you will find the information promised throughout the various tales of the book, plus some of Dan's and my favorite "go to" sources for information. Also see the resources listed in "Endnotes."

GENERAL

Back to Basics: How to Learn and Enjoy Traditional American Skills, by Readers Digest. Out of print but still available.

Craigslist for livestock, tools, and equipment: www.craigslist.org.

Encyclopedia of Country Living by Carla Emery. The classic how-to.

Forage Identification and Use Guide, University of Kentucky: http://www2.ca.uky.edu/agc/pubs/agr/agr175/agr175.htm.

Kinsey Agricultural Services, pasture soil testing with recommendations for organic soil ammendments. 297 County Highway 357, Charleston, Missouri 63834, (573) 683-3880, office@kinseyag.com, www.kinseyag.com/.

The Livestock Conservancy to learn about heritage breeds of livestock: www.livestockconservancy.org/.

State Cooperative Extension Offices for helpful local information: http://nifa.usda.gov/partners-and-extension-map.

YouTube for videos on how to do (or not do) just about everything: www.youtube.com.

CHICKENS

Henderson's Handy Dandy Chicken Chart, information about more than 60 chicken breeds: www.sagehenfarmlodi.com/chooks/chooks.html.

How To Preserve Eggs: freezing, pickling, dehydrating, larding, water glassing, & more from the eBook series *The Little Series of Homestead How-Tos from 5 Acres & A Dream* by Leigh Tate.

The New Chicken Coop. Blog series begins at http://www.5acresand adream.com/2014/02/the-master-plan-and-chicken-coop.html.

GOATS

Alpha chemicals for minerals for Coleby's Mineral Lick and citric acid for mozzarella cheese making, alphachemo8@yahoo.com, www.alphachemicals.com/.

"DIY Vitamins and Minerals for Goats," *5 Acres & a Dream The Blog*: www.5acresandadream.com/2013/03/diy-vitamins-minerals-for-goats.html.

Fias Co Farm, everything goat: http://fiascofarm.com/.

Herbal salves how-to at *5 Acres & A Dream The Blog*: www.5acres andadream.com/2010/12/herbal-salve-for-jasmine.html.

Kinder Goat Breeders Association: www.kindergoatbreeders.com/.

Onion Creek Ranch articles: www.tennesseemeatgoats.com/articles2/articlesMain.html.

Pearson Square how-to at *5 Acres & A Dream The Blog*: www.5acres andadream.com/2012/02/calculating-protein-with-pearson-square.html.

"NGS Geochemistry by County," *U.S. Geological Survey*, interactive soil mineral maps in the continental U.S. including selenium and copper. Useful for researching soil mineral deficiencies: http://mrdata.usgs.gov/geochem/doc/averages/countydata.htm

LLAMAS

Llama info: www.rockisland.com/~newmoon/llamas/llamainfo.html.

Llama-Training: www.llama-training.co.uk/index.html.

GUINEA FOWL

Frit's Farm Raising Guinea Fowl and Keets: www.guineafowl.com/fritsfarm/guineas/.

PIGS

American Guinea Hog Association: http://guineahogs.org/articles-about-guinea-hogs/.

"Weighing a Pig Without a Scale," *The Pig Site*: www.thepigsite.com/articles/541/weighing-a-pig-without-a-scale/.

HONEYBEES

Beekeeping for All, free download of Abbé Warré's book: www.users.callnetuk.com/~heaf/beekeeping_for_all.pdf.

"Plants That Bees Love" PDF downloads by Rusty Burlew at *Honey Bee Suite*: http://www.honeybeesuite.com/plant-lists/.

Pollen source by pollen color: https://en.wikipedia.org/wiki/Pollen_source.

warrebeekeeping Yahoo group: https://uk.groups.yahoo.com/neo/groups/warrebeekeeping.

ABOUT THE AUTHOR

LEIGH TATE and her husband Dan homestead five acres in the foothills of the southern Appalachian Mountains. Their goal is simpler, sustainable, more self-reliant living, and a return to agrarian values. In addition to critter keeping, gardening, food preservation, cheese making, and woodstove cookery, Leigh loves to write about homesteading. She is the author of the popular *5 Acres & A Dream The Book: The Challenges of Establishing a Self-Sufficient Homestead* and *The Little Series of Homestead How-Tos*. You can read about Leigh's and Dan's ongoing homesteading adventures at her blog, www.5acresandadream.com.

INDEX